Automation, work organisation and occupational stress

International Labour Office Geneva

ISBN 92-2-103866-1

First published 1984

ILO publications can be obtained through major booksellers or ILO local offices in many countries, or direct from ILO Publications, International Labour Office, CH-1211 Geneva 22, Switzerland. A catalogue or list of new publications will be sent free of charge from the above address.

Printed by the International Labour Office, Geneva, Switzerland

TABLE OF CONTENTS

Page

PART II

WORKING PAPERS

PREFACE

Much of the public concern about automation, and particularly the introduction of micro-electronic technology, has focused on the consequences for employment. However, besides its impact on employment levels, automation has far-reaching effects on the content of jobs, the structure of organisations and conditions of work. Thus, the effects of automation on work organisation and working conditions have also become the focus of attention in many countries.

A Meeting of Experts on Automation, Work Organisation, Work Intensity and Occupational Stress[1] was convened by the Governing Body of the International Labour Office in November-December 1983. This book contains the Report and Conclusions adopted by the Meeting and the working papers submitted to it as a basis for discussion.

The Report and Conclusions of the Meeting were noted by the ILO Governing Body at its February-March 1984 session. The Governing Body authorised the Director-General of the ILO:

(a) to distribute the report to the governments of ILO member States and, through them, to the national employers' and workers' organisations concerned; and

(b) to take account of the Conclusions of the Meeting when planning future ILO activities in this field.

The Conclusions begin with a summary of recent experience concerning automation and work organisation. They then turn to the effects on skills, worker responsibilities,

---

[1] A Meeting of Experts, in ILO terminology, is a meeting attended by experts appointed in their individual capacity by the Governing Body of the International Labour Office to advise the Governing Body and the Office on a well-defined technical subject.

careers and occupational stress; the implications for work organisation and job design; and the roles of governments and employers and workers and their organisations in the design of work with special reference to automation. Finally, they suggest future ILO activities in this field.

The following working papers were submitted to the Meeting and are reproduced in this publication: (1) Work organisation and the introduction of new technology: A survey of legislation and collective agreements in industrialised countries - prepared by the International Labour Office; (2) Automation and work organisation: Policies and practices in market economy countries - by B. Gustavsen; (3) Automation and work organisation: Policies and practices in countries with centrally planned economies - by L. Héthy; and (4) Work organisation and occupational stress - by C.L. Cooper.

PART I

REPORT AND CONCLUSIONS OF THE MEETING OF EXPERTS ON AUTOMATION, WORK ORGANISATION, WORK INTENSITY AND OCCUPATIONAL STRESS

REPORT

1. In accordance with the decision taken by the Governing Body of the International Labour Office at its 215th Session (February-March 1981) a Meeting of Experts on Automation, Work Organisation, Work Intensity and Occupational Stress was held in Geneva from 28 November to 7 December 1983.

Agenda of the Meeting

2. The agenda of the Meeting, as approved by the Governing Body at its 222nd Session (March 1983), was as follows:

(a) Recent experience concerning automation and work organisation in industry, commerce and offices:

  (i) impact on skills, worker responsibilities, careers and occupational stress;

  (ii) implications for work organisation and job design.

(b) The roles of governments and employers and workers and their organisations in the design of work with special reference to automation.

(c) Future ILO action.

Participants

3. Twelve experts were invited to attend the Meeting: four experts nominated by governments, four experts nominated by the Employers' group of the Governing Body and four experts nominated by its Workers' group. Three consultants recruited by the Office also attended as well as observers from the following organisations: International Organisation of Employers (IOE); International Confederation of Free Trade Unions (ICFTU); World Confederation of Labour (WCL); and World Federation of Trade Unions (WFTU). A list of the persons who took part is given at the end of this publication.

Opening of the Meeting

4. The Meeting was opened by Mr. V. Chkounaev, Assistant-Director-General, who welcomed the participants on

behalf of the Director-General. He noted that the Meeting reflected growing concern about the effects of rapid technological change on the nature of work. Accelerating automation was altering jobs and conditions of work in all sectors of economic activity. In some cases, the effects were positive: less strenuous work; a safer and healthier working environment; more meaningful and interesting jobs; better use of skills; and increased worker participation. The result was reduced fatigue and stress, higher motivation and greater production efficiency. In other cases, jobs resulting from advanced technology were repetitive, intense, socially isolated, low in skills and limited in scope for personal initiative. Occupational stress and its associated medical, psychological and behavioural problems were among the results. It had been shown that appropriate design of work could capture many of the advantages of automation and avoid most of the negative consequences . However, work design was complex and difficult. The Meeting could make an important contribution to knowledge about practical measures for overcoming these difficulties. The results of the Meeting would be submitted to the Governing Body.

Election of Officers

5. Mr. P.L. Rémy, Director of the National Agency for the Improvement of Working Conditions (France) was elected Chairman of the Meeting. After thanking the participants for their confidence in him, Mr. Rémy stated that the subject of the Meeting was of major importance for France and undoubtedly for other countries as well. It was now recognised that the worker must be taken into account in technological progress: the successful introduction of technology required appropriate implementation; and appropriate implementation required good conditions of work.

Presentation of the reports submitted to the Meeting

6. The Meeting had before it a background paper prepared by the Office as well as three background papers prepared by the consultants. These papers were briefly presented to the Meeting.

7. The working paper prepared by the Office entitled "Work organisation and the introduction of new technology: a survey of legislation and collective agreements in industrialised countries", was introduced by the Deputy Representative of the Director-General. She indicated that this working paper contained a brief survey of provisions

of legislation and collective agreements in a number of countries concerning work organisation and the introduction of new technology. The purpose of the paper was to illustrate differences in approaches, objectives, content and scope of these provisions. The extent to which legislation and collective agreements were used to deal with new technology varied according to the industrial relations systems and traditions of different countries. While collective bargaining could be an alternative to legislation, they were often complementary. The report described procedural provisions of legislation and collective agreements such as those on prior notification, consultation or negotiation and access to outside expertise. It also gave examples of substantive provisions on such issues as the content of jobs, work organisation, job design and the type of training or retraining for workers. Finally, she stated that legislation in centrally planned economies was supplemented by social development planning, models of work organisation and other practices such as work brigade systems.

8. Mr. C.L. Cooper introduced his paper on "Work organisation and occupational stress". The purpose of his paper was to highlight the social and psychological factors which are frequently sources of stress at work. Many of these factors take on greater significance during technological change. During the introduction of new technology, workers and managers were likely to experience occupational stress from the following sources: fear of losing their jobs; fear of impending changes in their job roles; and concern about being "untrainable" or deskilled. The main factor underlying these and other sources of stress was lack of control. This meant that workers' participation in the introduction of technological change was extremely important from the earliest planning through the process of introduction.

9. Mr. B. Gustavsen, in introducing his paper on "Automation and work organisation: policies and practices in market economy countries", stressed the challenges involved in developing institutions, policies and practical activities which resulted in concrete improvements. He noted that technological change had become sufficiently rapid that general theories were not particularly useful. Instead, it was necessary to develop policies and legislation that supported continuous, step-by-step improvements. In particular, enterprises needed support in developing more co-operative forms of work through, for example, reductions in the number of hierarchical levels of management and better integration at each level. This implied procedures for participation and negotiation which could deal with a broader range of issues more rapidly and more flexibly.

10. Mr. L. Héthy presented his paper on "Automation and work organisation: policies and practices in countries with centrally planned economies". He considered that technological change had led to positive and negative consequences in both centrally planned and market economies. The goal in both groups of countries was to reduce costs and increase benefits from these changes. However, the social, economic and political context of automation and work organisation influenced social consequences. Changes at shop-floor level needed support at higher levels of the organisation. It was therefore important to create a balance between central policies and decisions and local initiatives. In the centrally planned economies, improvements of the central institutions were relatively advanced, so that particular attention should be paid to supporting local initiatives.

Examination of the suggested points for discussion

11. The Meeting considered the organisation of its work on the basis of a list of suggested points for discussion which had been prepared by the Office. Several of the experts made general remarks or specific suggestions in this connection. The need to take economic realities into account was widely recognised. It was noted that the Meeting would need to consider both current technological changes and future developments. A working definition of automation would emerge from discussions of concrete reality, which would allow the diversity of national, sectoral and enterprise-level situations to be taken into account. The Meeting would emphasise practical means of making improvements.

12. The experts agreed that the suggested points for discussion would allow these general considerations to be taken into account. One specific suggestion was that the problems of occupational change should be added to the points for discussion in connection with the effects of changes in technology and job content on workers. With this modification, the suggested points were accepted as a basis for discussion.

Discussion of the first agenda item: Recent experience concerning automation and work organisation in industry, commerce and offices

13. The experts held an extensive discussion on the nature, extent and pace of technological change. It was

clear that there were considerable differences in the scope of change and the rapidity of its introduction.

14. Several examples suggested that automation in industry was often introduced more slowly than was popularly thought. Certain industrial technological changes required large investments and long periods of planning and construction. In addition, enterprises were often introducing automation in a planned rather than a piece-meal fashion. The process of implementation was also sometimes constrained in certain depressed sectors and areas by social and economic factors. Many small and medium enterprises had less access to new technologies and therefore introduced changes less rapidly, though this situation was changing. While employers were sometimes disappointed at the rate of introduction of techniques such as robotics, the need for competitiveness caused them to continue to make considerable efforts to automate.

15. In offices, commerce and services, the situation was somewhat different and the pace of change often was more rapid. Because investment per worker was lower, automated systems were more affordable. Office procedures and commercial record-keeping lent themselves to formalisation and programming. Examples were given by several experts of very rapid change through advances such as word processing, automatic tellers and check-out counters, computer-based stock control and new telecommunications equipment.

16. The diffusion of micro-electronics had been continuously encouraged by the decreasing size and cost of hardware. However, costs of software, the training of workers, the adaptation of workplaces, etc., had become an important proportion of the costs of automation. Moreover, the cost of much of the electro-mechanical equipment used in production processes remained high due to requirements for reliability and flexibility.

17. The experts agreed that the effects of the introduction of micro-electronic technology were not predetermined. There was a wide range of choice. Enterprises in the same industry used quite different systems to accomplish the same ends. Even in the automobile industry, for example, where the assembly line was once universal, modern factories exhibited considerable differences. These resulted in part from the increasing sophistication of production systems established to produce many variants on the same product. What mattered was not the technology itself but the way in which it was implemented. This in turn depended on the objectives of employers and trade unions and the procedures established for introduction of technological change. In

some cases, the objectives could be convergent, as in the desire to eliminate repetitive tasks.

18. It was clear that workers had become very apprehensive about the impact of new technologies on their jobs. In some countries, the current levels of unemployment added to their fears. Workers were also concerned about their ability to cope with the demands of sophisticated technological systems. At the same time, some experts stated that the social consequences of the introduction of new technology, for example with respect to employment security and worker attitudes, differed according to the socio-economic system concerned.

19. While workers' backgrounds and expectations were not the same, there were none the less certain needs that were relatively stable. Such needs as those for income, job security and communication with other workers were mentioned. The need for workers to have some control over their own job and work situation was particularly emphasised.

20. There was some disagreement as to whether suitable jobs would be available to satisfy these needs and expectations. Some experts felt that an adequate number of highly skilled jobs would be created while others disagreed. It seemed clear that highly educated workers might continue to have difficulties in finding jobs which met their expectations.

Impact on skills, worker responsibilities, careers and occupational stress

21. Several experts gave examples of cases in which workers had received advance information concerning new technology. They also referred to arrangements for the regular supply of information to workers. Such information helped workers to adapt to the new situation due to a better understanding of the production processes concerned, and had given them greater control over the content of their jobs.

22. It was generally agreed that automation had tended to eliminate certain particularly hazardous or arduous jobs. Several examples were given of the use of new technologies in this connection. On the other hand, some experts considered that new hazards were sometimes related to new technologies and that mental loads and occupational stress was becoming a more important problem than physical fatigue.

23. As regards the effects of new technology on the control workers exercised over their jobs, there were differing points of view. Some experts felt that increased worker responsibility would result from technological change. It was also mentioned, however, that there was a risk of a return to Taylorism and other monotonous forms of work organisation, and that the design of equipment could tend to dictate the pattern of work.

24. Some experts considered that new technologies were raising skill requirements and that a more highly skilled and versatile workforce would eventually be required. It was also pointed out that low-skilled workers were more easily displaced by new technology. One expert referred to the possible polarisation of skills: while a number of highly skilled jobs were created, new technology also resulted in a number of tedious and repetitive tasks.

25. One expert stated that, while the introduction of new technology brought greater flexibility, in some respects it also required greater flexibility on the part of the workers, for example, as regards the arrangement of working time. Several experts also drew attention to the problems of stress associated with the use of visual display units, noting that in some cases the number of hours permitted on terminals was restricted and the duration of rest breaks was increased. Another expert mentioned that new technology also created low-activity jobs, such as those of caretaker staff, who were required to be in attendance to ensure the efficient operation of high technology equipment should any problems arise.

26. Several experts also considered that new technology had an impact on career structures, pointing out that demand for some professions, such as graphic design, were disappearing while other professions were emerging. One expert noted that new job classifications and wage structures had developed for certain types of work.

27. The experts felt that it was necessary to attach particular importance to training. Workers should be trained for the jobs created by the introduction of new technology. It was also agreed that retraining should be made available to workers displaced by new technology, account being taken of the age and experience of the worker. It was noted, however, that some workers were unable or unwilling to take advantage of retraining courses offered. It was also noted that periodic retraining was likely to become a more regular feature of working life as a result of the changing demands of new technology. Several experts felt that older workers in particular required special consideration to help them adapt to technological change. Some experts also drew attention the problem of younger workers.

28. Several experts considered that the introduction of new technology could lead to interdependent tasks and group work, which increased communication and co-operation between workers. In some highly-automated plants, however, workers sometimes found themselves physically isolated. One expert noted that the use of work brigades in his country protected workers against isolation while another expert described the use of quality circles, which increased communication among workers. One expert also expressed concern that stress could be caused by the constant monitoring of workers' performance made possible by certain types of new technology. Some experts urged that particular care be taken concerning the security of data banks with a view to protecting the privacy of workers.

29. One expert felt it was important to concentrate on real and established problems rather than on theoretical difficulties. He considered that some of the problems under discussion fell into the latter category. Another expert considered that it was difficult to separate the many different aspects of the complex issue of new technology, and developments should be monitored using a multidisciplinary approach.

30. As regards the introduction of new technology in developing countries, one expert stated that workers and their representatives should be fully informed in advance of the introduction of new technology where this was not already being done. He also drew attention to the need to take social traditions into account in the industrialisation process.

Implications for work organisation and work design

31. Most speakers agreed that similar considerations had been important in the introduction of new technologies in both market-economy countries and those with planned economies. The reasons for the introduction of technology included increased productivity; the diversification and increased sophistication of products; the rationalisation of production and management; and the improvement in the quality of products. One expert noted that the conversion to automated systems could be time consuming and expensive. Another expert considered that the costs related to the implementation of new technology could greatly exceed the cost of the hardware. Particular attention was drawn to the problems faced by small and medium-scale enterprises which often lacked the resources and information necessary to cope with the effects of new technology.

32. Several experts referred to the role of management in the introduction of new technology, particular emphasis being placed on shop-floor supervisors and middle management. It was generally agreed that they often required training to help them adapt to their new responsibilities and to their new roles.

33. Another expert drew attention to the role that supervisory staff in developing countries could play in order to promote the gradual emergence of an industrial tradition. The question was raised whether developing countries could skip the Taylorist stage of industrial development. One expert recalled the desire of developing countries for economic independence and stressed that outdated technology should not be introduced in these countries. Experts also referred to the possibilities for improving co-operation between the industrialised and the developing countries in relation to the transfer of technology.

34. One expert gave examples of satisfactory economic results when appropriate account was taken of conditions of work. Another expert felt that such considerations would also help to limit industrial accidents, improve the industrial relations climate and reduce occupational stress.

35. Many experts also stressed the need to take full account of the effects of new technology on workers' health, both for economic and social reasons. A number of experts felt that consideration for workers' health must have priority over other considerations. One expert stressed that new technology should be used for humane purposes. He emphasised that hazards in the working environment could endanger the community as a whole. In this connection, he called for measures which would protect the international community from the misuse of advanced technologies.

36. One expert noted that in his country measures had been introduced to improve working conditions and to further the elimination of unpleasant tasks; in his view it was important to concentrate not only on occupational safety and health but also other aspects of working conditions. He felt that it was important to create compulsory standards regarding the introduction of new technology. Another expert felt that considerations concerning the impact of new technology needed to be made at the design stage by the manufacturer, and not later. Other experts, however, were opposed to the introduction of obligatory standards.

37. The importance of occupational medicine and the need for occupational physicians to develop special competence concerning new technology were referred to by several

experts. Occupational physicians should be consulted on the choice and design of new equipment.

38. Several experts felt that there was a possibility of introducing various forms of compensation for those workers who were most affected by occupational stress. Another expert suggested that rotation schemes could be used to share stressful tasks among several workers.

39. The arrangement of working time and the reduction of hours of work were also discussed as means of coping with the effects of new technology. One expert stressed the value of flexible working hours, while another referred to paid educational leave and the importance of leisure activities. Several experts also mentioned the possibilities of early retirement or special retirement arrangements for workers finding it difficult to adapt to the new circumstances.

40. One expert felt that it was unlikely that shift work would disappear as a result of the introduction of new technology; on the contrary the high cost of equipment and economic realities made it necessary to make the greatest possible and most cost-effective use of the new technology.

41. Several experts drew attention to the importance of effective measures for the provision of adequate and timely information to workers on all aspects of the introduction of new technology both before, during and on a regular basis after the implementation of new equipment and working methods. Some experts advocated the introduction of regulations governing the provision of such information. They considered that such information should include a description of the options available and the likely effects on the production process, work procedures and job content.

42. Several experts drew attention to the importance of decentralisation when consideration was being given to the likely effects of the introduction of new technology. One expert gave an example from his enterprise, in which responsibilities had been largely decentralised and problems were examined at the level of individual plants.

Discussion of the second agenda item: The roles of governments and employers and workers and their organisations in the design of work with special reference to automation

43. Several experts considered that many of the effects of the introduction of new technology on workers' health were still relatively little known. Governments

should therefore, in their view, finance research and studies on this subject. A number of experts referred to the situation in their countries as regards such research. Several experts considered that all competent disciplines should be mobilised to help increase understanding of the subject, emphasis being placed on both economic and social considerations. It was widely felt that research should be aimed at providing practical results.

44. Several experts referred to the role that governments should play in the dissemination of information on research results. They felt that governments should play a greater role in the dissemination of information, and that such information should be made widely accessible, particularly to small and medium-scale undertakings, which were often inadequately informed.

45. The responsibilities of governments in relation to initial and continuing training were highlighted by the experts. As regards initial training, several experts felt that it should provide general skills to help workers to adapt to innovation and new technology. One expert stressed that training should teach workers how to learn as well as what to learn. Other experts observed that education should be closely linked with employment opportunities. One expert stated that in his country training was planned on a medium-term basis in relation to human resources programmes in order to ensure that acquired skills were subsequently put to use at work. Several experts felt it was important to include information on protection against occupational hazards in order to raise awareness of the subject. As regards continuous training, it was generally agreed that this was a key to the successful introduction of new technology. Several experts believed that governments should take steps to ensure that retraining schemes were provided for workers whose jobs were displaced or modified by the introduction of new work processes. One expert felt it was important that workers should not live in fear of reductions in earnings, whatever the system used for the financing of retraining schemes. Another expert pointed out that difficulties were sometimes encountered in motivating workers to undergo retraining.

46. Several experts discussed the responsibilities of governments in relation to the training of specialists. New technology should be covered in the training of specialists in many different disciplines, such as medicine, engineering and social science, as it demanded a multidisciplinary approach.

47. There was a considerable range of views on the subject of legislation. One expert felt that legislation

should be kept to a minimum, as it was important not to encumber industry with excessive legislation which might endanger its economic viability. Another expert agreed, considering that collective bargaining was a more appropriate means of regulating the situation. Others argued that governments could not ignore the possible effects of the introduction of new technology, and that certain general regulations would have to be adopted by governments. One expert suggested that governments should produce sets of guidelines which would be useful in collective bargaining procedures.

48. The experts were unanimous in stressing the importance of the provision of adequate information to workers on the introduction of new technology. As to the exact nature and timing of such information, however, views differed. One expert felt that experience in his country had shown the advantages of presenting information in the form of a number of options. Another expert stressed that information should be provided both prior to and during the installation of technology which was not necessarily modern but which was new to the enterprise. The experts also stressed that information should be provided at the earliest possible opportunity not only to workers but also to engineers, safety experts and managers.

49. As regards the content of the information provided, it was suggested that it should include the aims of the undertaking and its economic situation, the type of equipment selected, the implications of new technology for job content and the likely effects on conditions of work. One expert stressed that the consultation and information phase should not neglect the shop-floor level staff. Another expert emphasised the importance of information and communication in creating a climate of confidence.

Discussion of the third agenda item: Future ILO action

50. In general, the experts considered that their earlier discussion had identified many potential areas for ILO action. However, owing to financial constraints within the ILO, it would be necessary to identify topics and priorities clearly. The ILO should focus its attention on action-oriented programmes and should not engage in theoretical research.

51. There was a consensus that the ILO should emphasise the collection and dissemination of information concerning the introduction of new technology: the ILO

should act as a clearing-house of information and should select, translate, summarise and disseminate information in a practical form as this would promote an exchange of experience among countries. One expert stressed the importance of disseminating information on the problems involved in the introduction of new technology. A number of experts felt that the ILO could draw to a greater extent on the experience of individual member States, including both centrally planned and market economy countries. Considerable amounts of information were available, and the ILO should select and disseminate the most useful data and research results on the subject. An added advantage would be that this might help prevent the duplication of research. In this connection one expert also observed that member States should take steps to ensure that relevant information was forwarded to the ILO.

52. Several experts considered that the wide range of issues associated with new technology and the variations among countries, technologies, sectors and sizes of enterprises made it impossible to elaborate international standards. Other experts disputed this argument and insisted that standards of a general nature could be developed to deal with such questions as the consultation of workers during the introduction of new technology. Another expert felt that the issues relating to new technology had not yet been adequately studied, and it was therefore too early to think of adopting standards; however, when more information was gathered on the subject it might be possible at that stage to take up the issue of standards.

53. Training and retraining was felt by a number of experts to be an important activity for the ILO. One expert referred specifically to the retraining of handicapped workers. Another expert suggested that the ILO should establish pilot programmes to provide training on the effects of the introduction of new technology. Other specific suggestions included activities in favour of small and medium enterprises, a study on home work and worker isolation and an in-depth study on new hazards and psychosomatic diseases.

54. In view of the fact that the problems posed for developing countries by the introduction of new technology could be quite different from those found in highly industrialised countries, one expert suggested that regional conferences should be organised to discuss various aspects of technological change.

Adoption of conclusions

55. The draft conclusions were examined and a number of amendments were made. The Conclusions were adopted unanimously.

CONCLUSIONS

Recent experience concerning automation and work organisation

1. The nature, extent and pace of technological change vary considerably in different countries, sectors and enterprises. In industry, where there are heavy investments in complicated automated systems, the pace of technological change may not be as rapid as has recently been the case in certain branches of the services sector. There are considerable differences as well according to the size of the enterprise.

2. Technological change has been more rapid than in the past in most cases and often very rapid. Nevertheless, there are differences in rates of change which are explained by two factors in particular:

- the frequently high cost of investments;

- the fact that modern automation often leads to a reorganisation of the production process, which makes the introduction of new technologies more complex.

This rate of change is likely to last for the foreseeable future. Most jobs will be significantly changed over the coming decades, some of them more than once.

3. New technologies can in the long run be positive for workers and for society, but they also carry risks.

4. There is a wide range of choices available when a new technology is designed and implemented. The national, economic, social and legislative context, the structure of the labour force, the objectives of the enterprise, management attitudes and practices and other factors influence these choices.

5. In the introduction of new technology, various social as well as economic objectives at both macro-economic and enterprise level should be taken into consideration.

6. Many different categories of workers are affected by technological change. Workers may be affected in differing ways. They are often apprehensive about their ability to cope with technological change, to obtain or retain a desirable job and to progress in their careers. Certain categories of workers, for example those above a certain age, may have difficulty in adapting to such change.

7. Worker expectations concerning the qualitative aspects of their jobs have been rising. Work is an important part of life for more than economic reasons.

8. Workers have specific needs in terms of job content. A particularly important need is for workers to have some control in this respect.

Impact on skills, worker responsibilities, careers and occupational stress

9. The introduction of new technology can affect all aspects of working life. It influences worker responsibilities, skill requirements, job content, physical and mental workload, career prospects, and communications and social relationships at work. In many cases, the effects in respect of these points depend as much on the way in which work is organised as on the technology itself.

10. The introduction of new technology can enhance worker responsibilities and the degree to which the worker has the discretion to make decisions concerning his job. On the other hand, the introduction of new technology can also be accompanied by fragmentation and specialisation of work and consequent reduction of the scope of this discretion.

11. New technology nearly always alters, to a greater or lesser extent, the nature of the skills required by jobs. In some cases, it has led to opportunities for skill development. In other cases, skill requirements have diminished considerably. Both of these effects may occur in the same enterprise. In such cases, there may be a risk that additional separations will be created within the workforce.

12. Automation frequently alters the nature of workloads. Many strenuous and hazardous tasks have been eliminated. At the same time, mental loads have tended to increase as a result of such factors as intensified pace or complexity of work.

13. Skill requirements and organisational patterns are not always compatible with the content of jobs resulting from new technologies.

14. The introduction of new technology can lead to inter-dependent tasks and group work. This assists in promoting communication and co-operation among workers. In some highly automated plants, however, workers may find themselves isolated. In other cases, communication may be

limited to impersonal contacts through terminals or similar devices.

15. Certain of the effects described above can contribute to occupational stress, which may be linked with certain serious health problems. An essential factor determining the degree of occupational stress is the extent to which the worker controls various aspects of his own job.

Implications for work organisation and job design

16. The introduction of new technologies can be a valuable opportunity to improve simultaneously productivity and conditions of work. Technological and managerial options are opening increasing possibilities for the design of work which takes both human factors and production efficiency into account. In order to generate positive effects on workers and minimise negative ones, the design of automated systems should take worker characteristics and preferences into consideration. These will vary from individual to individual and workplace to workplace. In addition, account should be taken of factors related to the technology itself, the economic situation, legislation and collective agreements and managerial attidudes and practices.

17. There is no universally desirable or effective work design. There are, on the other hand, certain techniques and procedures which are likely to lead to more successful implementation of new technologies.

18. An extremely important condition for success is the involvement of the workers themselves in the design of work. This has numerous advantages: it helps to assure worker acceptance of the technology and work procedures; it allows the workers' existing know-how to be put to use in the new production process; it promotes better worker knowledge of the new production process; it helps to motivate the workers; it provides workers with greater control over the content of their own work; and it aids in the successful implementation and smooth operation of the system.

19. In order for workers and their representatives to be fully involved in the design and implementation process, they need frequent and extensive access to information. This information should be provided in sufficient time for them to have significant influence on technologies and organisational choices. It should include a description of the technological options and the likely effects on production, procedures and work content.

20. The implications of new technologies require extensive action concerning all forms of training at national and enterprise level. Both access to training and the motivation to take advantage of training opportunities are important. Initial education should be designed to foster a capacity for lifelong learning.

21. Those workers displaced by automation will usually require retraining before they can find appropriate employment. The content and methodology of such retraining should take into consideration the age and experience of the worker.

22. Technological change means that most workers need access to continuing training. Support for such training is needed at both the national and the enterprise level.

23. Training required for the implementation of new technology should take place as early as possible. It should take account of existing know-how.

24. Enterprise-level training should emphasise the acquisition of multiple as well as basic skills. Work should be organised in ways which make use of such skills and which limit any tendency to retain large numbers of monotonous, unskilled jobs or to accentuate differences between highly skilled and unskilled jobs.

25. Workers and their representatives, including trade union officials, should receive training which will allow them to participate in the process of choice and implementation of new technologies.

26. Appropriate training and information should be provided to supervisors, systems designers, programmers and engineers concerning the likely effects of specific technological changes on workers.

27. Careful attention should be paid to conditions of work when new technologies are introduced.

28. The introduction of new technologies requires in the first place the application of general principles of occupational safety and health. Certain hazards are reduced by new technologies. These new technologies, however, can also generate new hazards relating to, for example:

- new forms of maintenance of equipment and machinery;

- reliability of processes;

- new toxic or dangerous substances; and

- radiation.

Aside from accidents, new technologies can also lead to work-related diseases, such as psychosomatic diseases. Such new hazards should therefore be carefully identified and their effects and methods of avoiding them should be studied.

29. Consideration of occupational safety and health should take place as early as possible in the planning and implementation of new technology, for example, through inclusion in specifications. In this connection, in addition to what has been stated above concerning workers and their organisations, the participation of occupational physicians, safety engineers and other specialists in work-related disciplines is required. Specific measures should be taken to ensure that they are able to carry out their role effectively.

30. The preventive role of occupational physicians should be expanded. They should increase their action concerning the working environment, especially during the design and implementation of new technologies. Occupational health services should promote preventive measures through the provision of advice and training to management, to workers and their organisations and to engineers and systems designers. Appropriate training should be provided to this effect during their education and throughout their careers. Regular medical examinations should be carried out and data should be carefully collected, especially concerning new hazards.

31. Because no system of preventive medicine can be perfectly effective, methods for the diagnosis and treatment of work-related diseases, particularly those related to new hazards, should be reinforced in the general health care system.

32. Orientation and on-the-job training for workers should include occupational safety and health. This subject should also be included in basic and vocational education and training where appropriate.

33. The level and the organisation of working time are an important element to be taken into account in the introduction of new technologies. Owing to the cost of equipment, the hours during which it is used may need to be prolonged, but automatic operation can sometimes reduce the level of manning during certain periods. For certain

tiring and stressful tasks, workers should benefit from breaks, shorter hours or other alleviating measures. These considerations may lead to modifications in the arrangement and length of working time. Attention should be given to the disadvantages liable to be associated with shift work, night work and changes in the scheduling and length of working hours. Such disadvantages can be attenuated if overall hours of work are reduced.

34. Workers who are likely to experience particular problems because of the introduction of new technology should benefit from special attention in the design of technological systems. They could also benefit from special measures concerning their conditions of work.

35. Middle managers have special responsibilities concerning the introduction of new technology. They are familiar with shop-floor reality. At the same time, their own jobs are often altered. They therefore require training concerning the implications of new technologies for workers and for their own roles.

36. The introduction of new technology can be an opportunity to improve the quality of working life and to avoid excessive stress. In this connection, job design techniques can permit, in particular, an increase in the control of workers over their jobs; avoidance of excessive pace of monotony or work; expanded opportunities for communications and co-operation on the job; and the development and fuller use of skills.

Roles of governments and employers and workers and their organisations in the design of work with special reference to automation

37. The preceding analysis has shown the diversity and complexity of the problems which new technologies pose concerning conditions of work and work organisation. Employers and workers and their organisations as well as governments will have to take these multiple factors into account in their policies, activities and negotiations.

38. It is clear that the concrete ways these factors are taken into account will vary considerably from country to country. The level of development, the organisation of the economic system, labour relations law and practice, social attitudes and other factors influence the respective

roles of governments, employers and workers and the most effective means of action available to them.

39. Action concerning both new technologies and the design of work involves the joint responsibilities and interests of governments, employers and workers. Co-operation among all the parties concerned should therefore be reinforced at all levels. Particularly promising methods to take advantage of the realism and practical expertise of the different parties include use of mechanisms for prior consultation during the development by States of legislation and policy concerning technology and working life. Along the same lines, the participation of these different parties is desirable:

- in councils or other bodies which advise or supervise institutions working in these fields;

- in the definition of research objectives and the dissemination of research results;

- in the development of educational, training and public information activities in these fields.

40. It is especially important that the introduction of new technology take place in an atmosphere of credibility and mutual trust between management and workers. This can be promoted by regular consultation concerning new technology; by active communication programmes for top management, middle management and workers; by orientation and training programmes, especially those carried out jointly by management and workers' representatives; and by joint identification and analysis of technological options as early as possible.

41. The effects of new technology on workers are complex and difficult to analyse. The development of concrete improvements must take into account any technological, economic and legal factors in addition to social criteria. It is therefore important to make integrated use of all relevant disciplines in this process, and in particular occupational medicine, safety engineering and ergonomics, industrial and civil engineering, labour law, occupational psychology and management sciences.

42. The development and application of new technology involves many different enterprises, institutions and groups. In order to take better account of the problems posed by new technology for conditions of work and work organisation, communication and joint action should be developed among the manufacturers and suppliers of hard-

ware and software; institutions and individuals conducting relevant basic and applied research; employers' and workers' organisations; the enterprises concerned; and experts who intervene at the enterprise level, particularly occupational physicians. All persons involved in the design, application and operation of automated systems should be made aware of the needs of the users of the systems.

43. The introduction of new technology may require the re-examination of standards concerning occupational safety and health and conditions of work to ensure that essential protection is provided to workers. In some cases, the revision of existing standards or the development of new ones may be necessary. However, standards sometimes do not keep pace with the rate of technological progress. In addition, it has proved difficult to establish standards concerning some aspects of working life, such as occupational stress, because of problems of identification and measurement. Moreover, excessively rigid or detailed standards can be difficult for enterprises to apply and for inspectors to enforce.

44. In conditions of rapid technological change, a promising approach has been to emphasise in national legislation enterprise-level procedures for improving the working environment. Such procedures call for the participation of workers and their representatives in developing and applying improvements, in particular during the introduction of new technology. This approach permits greater flexibility in the design of improvements adapted to the characteristics of the technology and the workplace and the needs and priorities of the workers concerned.

45. Collective bargaining and consultation with workers' representatives are particularly effective means of dealing with issues related to the introduction of new technologies because they can take into account the interests of those most directly concerned as well as the specific conditions at the workplace. Collective agreements may, where appropriate, and particularly where these questions are not regulated by law, include provisions relating to procedural issues such as the timing for, and content of,information to be provided to workers and their representatives; requirements for consultation and negotiation when new technology is introduced; training of workers and their representatives; and access to outside expertise concerning technical specifications and options and work design procedures. Collective agreements may also refer to substantive issues such as: the training or retraining of workers whose jobs have been altered or

eliminated; the job characteristics which should be sought during the introduction of new technology; and questions relating to supervision, work measurement and job evaluation, skill and job classification, remuneration and job security, and hours of work and the organisation of working time. Where appropriate, national or sector framework agreements may be used to promote and assist the development of enterprise- and plant-level agreements.

46. Both national and enterprise level responsibilities in the provision of training are extensive. Training is required for all those who may be concerned with the design of work. The content of such training should include the social as well as the economic implications of technological change. Practical training and information materials for these purposes are a particular priority. Such materials should be developed with the participation of representatives of employers and workers. Access to appropriate training for workers' representatives for the purpose of studying the implications of new technologies should be encouraged.

47. National systems for both general and vocational and technical training should be examined with a view to ensuring that the skills required by new technologies are adequately represented in the labour force. At the same time, education and training should encourage the development of managerial skills and orientations which reflect the new social and technological situation in workplaces.

48. Research must keep pace with the effects of technological change at shop-floor level. Tripartite participation and multidisciplinarity in research activities should be reinforced. Governments have a particular responsibility to support research and to promote co-ordination among the many different institutions involved.

49. The following research topics are among those with high priority:

- collection and analysis of data on the hazards which may be related to the introduction of new technologies, their causes and their effects;
- identification of effective means of prevention and treatment of stress;
- case-studies which elucidate the reasons for both positive and negative experience with the introduction of new technology at shop-floor level;

- identification of technological and organisational options on an industrial or sectoral basis which permit consideration of social factors during the design of work; and

- the study of management techniques which are adapted to changing technology and worker expectations.

50. Research should be practically oriented. Care should be taken to ensure that the end products of research are comprehensible to, and usable by, the persons most concerned with work design. Solutions which are accessible without recourse to sophisticated expert advice should be emphasised.

51. The dissemination of information on the effects of the introduction of new technology is important in order to enable governments, employers and workers and other concerned to take advantage of the opportunities offered by new technology to improve productivity and conditions of work and to avoid or minimise any negative effects.

52. There is a particular need for information on:

- technological developments and their applications;

- research findings concerning effects on safety, health and the working environment and measures for prevention, diagnosis and treatment;

- techniques of work organisation, work design and ergonomics adapted to new technology and to worker needs and expectations;

- the provisions of laws, regulations and collective agreements.

53. Appropriate information should be available to government officials, employers and workers and their organisations, manufacturers and suppliers, specialists in fields relating both to technology and to various aspects of working life, and the general public.

54. There is a need to select high quality and relevant information from the large volume available and to ensure that it is accessible to those concerned.

55. The following measures may be useful in this connection:

- establishment of data banks containing information on

technology and working life and on foreseeable developments in these fields;

- use of advanced information technology to provide access to such data;
- establishment of linkages among research institutes, enterprises, employers' organisations, trade unions and other sources and users of information in this field;
- support for the adaptation of research results into forms which can be directly used at the enterprise level.

Future ILO action

56. The implications of new technology for conditions of work and work organisation are and will continue to be of critical importance for governments, employers and workers and for society as a whole. Many issues within this field will undoubtedly require the attention of the ILO. The extent and nature of ILO activitiy in this field will depend on the overall priorities of the Organisation. Emphasis should be placed on areas and means of action in respect of which the ILO is well placed to make a useful contribution.

57. Given the pace and complexity of technological change and the diversity of national conditions, the development of international labour standards in this field appeared premature to certain experts, at least for the near future. However, in so far as a consensus may emerge on such general principles as the consultation of workers and their representatives concerning the introduction of new technology, the possibility of considering international labour standards in this field should be re-examined.

58. Rather than undertaking research of an academic or theoretical nature, the ILO should concentrate on studies with a clearly practical orientation. Such studies should provide comparative information on experience in different regions, countries and sectors. They should describe trends and developments in relation to specific problems and the various approaches being used to secure improvements, including legislation, collective agreements and enterprise level techniques.

59. The ILO should particularly emphasise the collection, analysis and dissemination of information on the implications of new technology for conditions of work and work organisation. It should take into consideration for this purpose the information needs of various government agencies, employers and workers and their organisations, research institutions and others concerned.

60. Priority should be given to information which has a practical orientation. The ILO should make a special effort to facilitate exchange of information among different countries and regions. Where possible, the ILO should use its facilities to overcome difficulties in information exchange due to language.

61. The ILO's function as a clearing-house for information should include the promotion of linkages between institutions concerned with new technology as it relates to conditions of work and work organisation; the identification of information sources in various countries; the acquisition of information of particular international relevance; the selection, processing, storage and active dissemination of such information. Where appropriate, advanced information technology should be used in carrying out these functions. Given the large and rapidly growing volume of information, the ILO should devote particular efforts to the selection and dissemination of authoritative information. Where necessary, this information should be converted into simplified or summarised form.

62. Because of the importance of training in relation to the introduction of new technology and the organisation of work, the ILO should develop practical materials for use in such training activities as workers' education programmes, management development programmes and vocational training and rehabilitation projects. Appropriate materials should also be developed for safety delegates, labour inspectors and other personnel directly involved in the protection of workers.

63. These Conclusions relate primarily to problems of concern to industrialised countries, although several may also be applicable to developing countries. Other specific questions merit further study at regional meetings of these countries, in particular those related to transfer of technology and the training of management and workers. It is necessary, however, to emphasise that a genuine transfer of technology is necessary in order to allow the developing countries both to overcome the problems related to the introduction of new technology and to provide training

for national managers and highly skilled workers. Such a transfer of technology would also allow these countries to carry out basic and applied research with a view to the adaptation of technology to their particular circumstances.

PART II

WORKING PAPERS

WORK ORGANISATION AND THE INTRODUCTION OF NEW TECHNOLOGY

A Survey of Legislation and Collective Agreements in Industrialised Countries

International Labour Office

Introduction

1. At its 215th Session (February-March 1981), the Governing Body of the International Labour Office decided to convene a Meeting of Experts on Automation, Work Organisation, Work Intensity and Occupational Stress. At its 222nd Session (March 1983), it decided to hold the meeting in Geneva from 28 November to 7 December 1983, and to fix the agenda as follows:

(a) Recent experience concerning automation and work organisation in industry, commerce and offices:

  (i) impact on skills, worker responsibilities, careers and occupational stress;

  (ii) implications for work organisation and job design.

(b) The roles of governments and employers and workers and their organisations in the design of work with special reference to automation.

(c) Future ILO action.

2. The meeting will have before it four working papers. Three of these papers, prepared by ILO consultants, deal with particular aspects of the theme of the meeting: work organisation and occupational stress; automation and work organisation: policies and practices in market economy countries; and automation and work organisation: policies and practices in countries with centrally planned economies. The present paper, prepared by the International Labour Office, contains a brief survey of provisions of legislation and collective agreements in a number of industrialised countries relating to the implications for work organisation of the introduction of new technology.

3. The introduction of new technology has led to extensive changes in the organisation of work. These

changes have affected the content of jobs, the relationships between different jobs and the conditions under which the jobs are performed. They have affected the division of labour, job requirements and procedures, skill and training requirements, career opportunities, inter-personal relationships at work, organisational structures and the roles of managers and supervisors and workers and their representatives.

4. Many of the effects of new technology are not simply consequences of the technology itself. A greater variety of technological alternatives and more flexible technology can provide a wider range of choices and thus more scope for improved work organisation from the workers' perspective. But the reasons for introducing the technology, the tasks for which the technology will be used and the way in which the work is then organised are often the critical factors in determining the effects the technology will have. Different variations of new technology can be developed with different applications, forms of work organisation, or types of management in mind.

5. In addition, conditions outside the workplace or the individual firm influence decisions concerning new technology. The political, economic and social context in which technology is introduced cannot be overlooked.

6. The issues that arise in connection with the introduction of new technology can be both procedural or substantive. Procedural issues can relate, for example, to the following points:

- provision of information to workers and their representatives;
- a requirement for consultation or negotiation;
- the establishment of procedures for such consultation or negotiation;
- access to outside expertise by workers and their representatives;
- training for workers' representatives.

Examples of substantive issues include the following:

- questions relating to security of employment;
- training or retraining for workers whose jobs are eliminated or changed;

- questions relating to work pace, supervision, fragmentation of jobs, downgrading;

- questions relating to health, safety and ergonomics;

- questions relating to remuneration.

7. Both procedural and substantive issues are dealt with in different countries through legislation, through collective agreements or through a combination of the two. This paper contains a limited survey of the main provisions of legislation and collective agreements available in the Office relating to the introduction and use of new technology with respect to the subject of the meeting. The first part of this paper covers industrialised market economy countries and the second part, countries with centrally planned economies. The survey has of necessity been limited to countries on which sufficient information was available at the time of writing.

8. The purpose of this paper is to assist the meeting in its discussions. The presentation is meant to be illustrative rather than exhaustive. Examples are given of provisions of legislation and collective agreements to show differences in procedures, content and scope. The inclusion of certain provisions does not imply that they are commonly found in legislation or collective agreements: the aim has been to cite a wide range of examples in order to illustrate different approaches.

I. Provisions of legislation and collective agreements in selected industrialised market economy countries

9. The extent to which legislation or collective bargaining or both are used to deal with issues relating to the introduction of new technology and work organisation varies according to the industrial relations systems and traditions of different countries.

10. All countries, of course, have at least some laws and regulations - for example, those on safety and health, conditions of work, and training - which pertain directly or indirectly to such issues. The legislative provisions cited in this part of the paper are explicitly concerned with either general or specific aspects of technology and work organisation.

11. These provisions can be divided into two broad groups. The first includes legislation aimed essentially at establishing rules, procedures or machinery for the involvement of workers or their representatives in the consideration of technology-related issues. Legislation of this kind can cover such points as the rights and obligations of employers and workers; areas of consultation, negotiation or co-determination; and mechanisms for resolving conflicts. The second group includes legislation which, either directly or through the granting of regulatory authority, addresses substantive problems. The objectives can range from the establishment of general criteria or guidelines for job design to the promulgation of specific standards concerning particular types of technology or aspects of the working environment.

12. Whether serving mainly as a complement or as an alternative to legislation, collective bargaining also deals with both procedural and substantive measures. Collective agreements have been concluded at various levels: national, sectoral, company and plant. The scope of collective bargaining has been expanding to cover subjects which previously tended not to be open to negotiation. An important and relatively recent development has been the conclusion of agreements dealing exclusively with the introduction of new technology, commonly referred to as "new technology agreements". In some countries, legislation has established the obligation to negotiate on certain matters; in others, the rights acquired through collective bargaining have been consolidated and generalised through legislation.

13. In this part of the paper, relevant provisions of a number of legislative texts and collective agreements are presented. It must be emphasised again that the survey is not meant to be complete or exhaustive. An effort was made to give examples of provisions to show differences in objectives, content, procedures or coverage with respect to the issues within the scope of the meeting.

A. Legislation

14. In this section, examples are given of legislative or other statutory provisions relating to some of the specific issues mentioned above. These include the provision of information; consultation and negotiation; co-determination; access to outside expertise; training for workers' representatives; job content, work organisation and job design; training for workers whose jobs have been changed or eliminated; and regulations concerning visual display units (VDUs). This is again not a comprehensive survey: many of the countries covered have more extensive provisions than those cited. The purpose here is to illustrate different approaches.

Provision of information

15. In the Federal Republic of Germany, the Works Constitution Act of 1972[1] contains detailed requirements as to the provision of information to works councils.

General

- "The employer shall supply comprehensive information to the works council in good time to enable it to discharge its duties under this Act. The works council shall, if it so requests, be granted access at any time to any documentation it may require for the discharge of its duties; ... shall be entitled to inspect the payroll showing gross wages and salaries of employees." [Article 80(2)]

Prospective changes in company structures, design of jobs, operations and organisation of work

- "The employer shall inform the works council in due time of any plans concerning:

1. the construction, alteration or extension of works, offices and other premises belonging to the establishment;

2. technical plant;

3. working process and operations; or

4. jobs ...". [Article 90]

- The employer shall inform the works council "in full and good time of any proposed alterations which may entail substantial prejudice to the staff ...". These alterations include reduction of operation, closures, mergers or transfers of departments or the establishment, important changes in the organisation, purpose or plant of the establishment and the "introduction of entirely new work methods and production processes". [Article 111]

Manpower planning

- The employer shall inform the works council in "full and in good time of matters relating to present and future manpower needs and the resulting staff movements and vocational training measures and supply the relevant documentation ...". [Article 92]

- The employer shall inform the works council "in advance" of any individual staff movements - "engagement, grading, regrading and transfer" - and submit to it the appropriate documentation. He shall also inform the works council of the "implications of the action envisaged ... and obtain its consent to the action envisaged". [Article 99]

- The employer shall also notify the works council of any temporary staff movements [Article 100] and the works council will be consulted on any dismissal [Article 102].

Information rights of the Finance Committee*

- The employer shall inform the Finance Committee "in full and in good time of the financial affairs of the

---

* The Finance Committee is a committee of the works councils. Its duty is to consult with the employer on financial matters and report to the works council.

establishment and supply the relevant documentation ...". The financial matters covered include among others "production and investment programmes", "rationalisation plans", "production techniques and work methods, especially the introduction of new work methods". [Article 106]

16. In the Netherlands, the Works Councils Act of 1979 makes a distinction between information which the employer is obliged to provide on request and information that must be provided in any case.

- "An employer shall, on request, promptly provide the works council and its committees with all the information they reasonably need for the performance of their duties ...". [Article 31]

- The employer must provide detailed information concerning the economic and legal structure of the establishment [Article 31(2)]; past and prospective financial status of the company [Article 31a]; labour statistics tracing the effects of the employer's social policy as well as forecasts of staff movements and social policy [Article 31b]; and the employer's intention to seek the advice of an outside expert on specified social policy matters [Article 31c].

17. In Norway, the Work Environment Act of 1977 states:

- "Workers and their elected representatives shall be kept informed of the systems used for the planning and execution of work, including any projected alterations to such systems". [Article 12(3]

18. In Sweden, the Act respecting co-determination at work of 1976 states:

- "An employer shall keep a worker's organisation in relation to which he is bound by a collective agreement continuously informed of developments in the production and financial aspects of his business and also of the principles on which his personnel policy is based. He shall likewise afford the organisation an opportunity of examining any books, accounts and other documents relating to his business, to the extent that the organisation requires to do so to safeguard its members' joint interests in relation to him". [Article 19]

19. In Italy, Law No. 833 of 1978, which establishes the National Health Service, requires the employer to give information on the production process: plant, equipment and machinery and materials and substances used in production or resulting from production.

20. In the United Kingdom, two Acts provide for information disclosure:

- The Employment Protection Act of 1975 requires the employer to disclose to trade union representatives information required for "the purposes of collective bargaining". [Article 17]

- The Health and Safety at Work Act 1975 gives workers and their representatives the right to any information necessary for their health and safety. Safety representatives must be supplied with information concerning any plans or changes which affect the health and safety of workers.

21. In France, Act No. 82-915 of 28 October 1982 concerning the development of institutions for employee representation gives works councils greater access to specific written information on company operations. A group council system will be established between parent companies and affiliates to co-ordinate and analyse information. Group councils are entitled to receive information on the company's activities, financial situation, and employment trends, both on a group basis and in respect of its dependent enterprises.

Consultation and negotiation

22. In the Netherlands, the Works Council Act provides that a works council "shall be afforded an opportunity by the employer of expressing its opinion on any decision he intends to take" in connection with, among others, changes in the activities and organisation, recruitment, major investments, hiring an outside expert, major contracts with other enterprises, etc. [Article 25]. The Working Environment Act of 1980 states:

- "The employer shall hold prior consultations on the policy of the undertaking in so far as it may have a demonstrable effect on the safety, health and welfare of the workers ...". [Article 4(4)]

23. In France, the Labour Code as amended on 20 January 1981 states:

- "The works committee shall accordingly be consulted in all cases before new methods for the organisation of work are introduced, before any major change is made to work stations as a result of alterations to equipment or the organisation of work, before any alterations are made to rates of working or standards of productivity; whether or not they are linked to the remuneration paid, and before any major change affecting the environmental conditions or occupational safety is introduced". [L.437-1]

Act No. 82-689 of 4 August 1982 on employees' rights to express their views also provides the following:

- "... the employees shall have the right to express their views, directly and collectively, on the content and organisation of their work and on the conception and implementation of schemes for improving working conditions within the undertaking". [L.461-1]

24. The Act respecting co-determination at work of 1976 of Sweden states:

- "Before an employer decides on any important change in his activity, he shall on his own initiative negotiate with the workers' organisation in relation to which he is bound by a collective agreement. The same shall apply before an employer decides on any important change in the working conditions or conditions of employment of workers ...". [Article 11]

In addition, the Act provides that the workers' organisation can request the employer to open negotiation in cases other than those referred to in Article 11 before the employer takes or implements a decision affecting members of the organisation. [Article 12]

Co-determination

25. The Works Constitution Act of 1972 of the Federal Republic of Germany gives the works council co-determination rights on certain matters as far as they are not prescribed by legislation or collective agreement. These matters include:

- "the commencement and termination of the daily working hours including breaks and the distribution of working hours among the days of the week";

- "any temporary reduction or extension of the hours normally worked in the establishment";

- "the introduction and use of technical devices designed to monitor the behaviour or performance of the employees";

- "arrangements for the prevention of employment accidents and occupational diseases and for the protection of health on the basis of legislation or safety regulations";

- "questions related to remuneration arrangements in the establishment, including in particular the establishment of principles of remuneration and the introduction and application of new remuneration methods or modification of existing methods";

- "the fixing of job and bonus rates and comparable performance-related remuneration including cash coefficients (i.e. prices per time unit)". [Article 87(1)]

26. The Works Councils Act of 1979 of the Netherlands states:

- "An employer shall require the consent of the works council to any decision he intends to take in connection with the introduction, amendment or cancellation of -

  ...

  (c) any arrangements for hours of work or leave ...;

  (d) any system of remuneration or job assessment;

  (e) any arrangements in connection with occupational safety, health and welfare;

  (f) any arrangements in connection with recruitment, dismissal or promotion policy;

  (g) any arrangements in connection with staff training;

  (h) any arrangements in connection with the assessment of staff;

  (i) any arrangements in connection with works welfare services;

(j) any arrangements in connection with consultations relating to work; ...". [Article 27(1)]

An analogous provision is found in the Working Environment Act of 1980 in Article 14(6) concerning the obligation of the employer to the working environment committee.*

Access to outside expertise

27. The Works Constitution Act of 1972 of the Federal Republic of Germany provides the following:

- "In discharging its duties the works council may, after making a more detailed agreement with the employer, call on the advice of experts in as far as the proper discharge of its duties so requires". [Article 80(3)]

28. In the Netherlands, the Works Council Act of 1979 gives the works council the "power" to invite "one or more experts or members of the managerial staff to attend its meetings or submit papers" [Article 16]. In addition, expenses incurred by the works council and its committees in connection with expert advice obtained shall be borne by the employer if he was previously informed about it, subject to a decision by a joint inter-enterprise body known as the trade commission if the employer objects to covering the expense [Article 22]. The Working Environment Act of 1980 also provides for similar rights for the working environment committees [Article 11].

Training for workers' representatives

29. In Denmark, the Act respecting the Working Environment of 1975 requires the employer to provide members of the safety committee with "an opportunity for acquiring the necessary knowledge of or training in safety questions" [Article 6(3)]. This obligation is important with respect to Article 6(4) which requires employers to give the safety committee an opportunity to take part in planning in so far as they relate to safety and health at the workplace.

---

* In the Netherlands, the working environment committee consists only of representatives of the workers. In Norway, the working environment committee includes representatives from the employer, workers and the safety and health personnel.

30. The Working Environment Act of 1980 of the Netherlands also states that members of the working environment committee should "be afforded the opportunity, during working hours and on full pay, of receiving such education and general training as they reasonably require for the discharge of their duties". [Article 14(10)]

Job content, work organisation and job design

31. Although still relatively rare, legislation has also been used directly to promote improvements in job content, job design and work organisation. Such legislation is based on increasing evidence that a job can be inappropriate or damaging to the worker without there being any obvious immediate physical harm. Accordingly, the legislation goes beyond the protection of workers' physical well-being to include consideration of social and psychological factors which should influence the content and design of jobs and workplaces.

32. For example, the legislation in Norway, entitled "Act respecting workers' protection and the working environment" (Work Environment Act of 1977), states:

- "The purpose of this Act is to -

  1. ensure a working environment that provides workers with complete safety against physical and mental hazards and with a standard of technical protection, occupational hygiene and welfare corresponding at all times to the technological and social progress of society;

  2. ensure safe working conditions and a meaningful employment situation for the individual worker ...". [Article 1]

- "The working environment in an undertaking shall be entirely satisfactory from the standpoint of both an individual and an overall assessment of the environmental factors that are capable of having an effect on the workers' physical and mental health and welfare". [Article 7]

Section 12[2] of this Act contains extensive provisions concerning job design and work organisation. These provisions are based on the idea that there is interdependence between job quality and systematic efforts to resolve work environment problems at enterprise level.

The relevant provisions of Section 12 are as follows:

- "(1) General requirements. The technology, organisation of work, timetables and systems of remuneration shall be so conceived that the workers are not exposed to adverse physical or mental effects or that their chances of devoting care and attention to safety considerations are reduced.

  Conditions shall be so designed that the workers are afforded a reasonable opportunity of occupational and personal advancement through their work.

  (2) Conception of work. Account shall be taken, in the planning and conception of work, for the opportunities enjoyed by the individual worker for taking his own decisions and assuming responsibility for what he does.

  An effort shall be made to avoid monotonous repetitive jobs and jobs which are determined by a machine or conveyor belt to such an extent that the worker is prevented from altering his rate of working.

  An attempt shall be made in other respects to conceive work in such a way that there are opportunities for variety and contact with other persons, for establishing a relationship between individual tasks and for the workers to keep themselves informed of production requirements and results.

  (3) Special provisions as to systems of management and planning. Workers and their elected representatives shall be kept informed of the systems used for the planning and execution of work, including any projected alterations to such systems. They shall be given such training as they need to become acquainted with such systems and shall take part in their conception.

  (4) Special provisions as to work involving hazards. Systems of payment by results shall not be used in work where they may have a material influence on safety ...".

Another important aspect of the Norwegian legislation is that it places strong emphasis on workers themselves to take an active role in investigating and improving their workplaces [Article 16]. This is in contrast to other legislation where the initiative is with the employer.

33. In Sweden, the Working Environment Act of 1977 envisages similar goals:

- "(1) The working environment shall be satisfactory, having regard to the nature of the work and the social and technical progress of society.

  Working conditions shall be adapted to human aptitudes. An effort shall be made to arrange the work in such a way that an employee can himself influence his work station.

  (2) Work shall be planned and arranged in such a way that it can be done in a healthy and safe environment". [Chapter 2, Articles 1 and 2]

34. In the Federal Republic of Germany, the Works Constitution Act of 1972 has two important provisions relating to job design and work organisation.

- In their consultations concerning any plans relating to construction, alteration or extension of works, plant, working process or jobs, "the employer and the works council shall have regard to the established findings of ergonomics relating to the tailoring of jobs to meet human requirements." [Article 90]

- "Where a special burden is imposed on the employees as a result of changes in jobs, operations or the working environment that are in obvious contradiction to the established findings of ergonomics relating to the tailoring of jobs to meet human requirements, the works council may request appropriate action to obviate, relieve or compensate for the additional stress thus imposed ...". [Article 91]

A notable feature of this legislation is its emphasis on a scientific basis for improvements in job design, job content and work organisation.

35. In the Netherlands, the Working Environment Act of 1980 contains provisions concerning the characteristics of jobs and measures to be taken to improve job content and work organisation which the employer must take into account. These provisions have some similarities with the Works Constitution Act of 1972 of the Federal Republic of Germany and the Working Environment Act of 1977 of Norway. The relevant text states:

- "An employer shall take account of the following considerations when organising the work, installing the workplaces and determining the methods of production and work:

(a) such methods of production and work must be adopted and such measures taken (which shall include the provision of appropriate and suitable means of protection for the workers) as will, in the light of the best existing principles of technology, the current state of industrial health care and the current state of knowledge in the field of ergonomics and industrial sociology, ensure the greatest possible degree of safety, the greatest possible degree of health protection and the promotion of the greatest possible attention to the worker's welfare, unless the foregoing cannot reasonably be required;

...

(e) the installation of workplaces, the working methods employed and the aids used for the purposes of the work must be adapted to the worker; where this cannot be achieved or cannot be satisfactorily achieved, the work must be interrupted by regular breaks or varied by the performance of other work;

(f) monotonous work involving repetitions at short intervals and work in which the rate of working is determined in such a manner by a machine or conveyor belt that the worker himself is prevented from influencing the rate of working must be avoided as far as can reasonably be required;

(g) when the various jobs are conceived and allocated, account must be taken of the worker's personal characteristics, such as age, sex, physical and mental constitution, experience, skills and knowledge of the working language; whenever such characteristics so warrant, due account also being taken of the circumstances of the work, special supervision must be exercised or other suitable arrangements made to promote the safety and protect the health of the workers concerned;

(h) unless he cannot reasonably be required to do so, the employer must organise the work in such a way, having regard to the nature of the establishment and the circumstances of the work, that the work does not involve any adverse effects for the worker's physical and mental health;

(i) in so far as can reasonably be required, the content of a worker's job must be so determined, and the co-operation, management and supervision that the work involves be so arranged, that the work contributes to the worker's personal development and the improvement of his skills and that sufficient opportunities are provided for him to arrange the work in accordance with his own conception of it (as also determined by his skills), to maintain contact with other workers and to keep himself informed of the purpose and result of what he does and the requirements laid down for its performance". [Article 3(1)]

Training for workers whose jobs have been changed or eliminated

36. The legislation examined reveals differences in statutory provisions concerning training for workers. This legislation does not refer explicitly to the introduction and use of new technology. However, in many cases, it is used by workers and their representatives to obtain training and/or retraining necessitated by changes in their jobs during the application of new technology. Statutory provisions vary with respect to their purpose, content and target groups. Some provide for general training to enable workers to participate in the design of their jobs; others to enable workers to adapt and perform their jobs efficiently and safely. In some cases, legislation also specifies the type of information to be given to workers as part of general training.

37. As previously mentioned, the Norwegian Working Environment Act of 1977 specifies that workers shall be given training "as they need to be acquainted" with systems used for planning and execution of work so that they can "take part in their conception". [Article 12(3)]

38. In the Netherlands, the Working Environment Act of 1980 provides for general training for young workers. In this connection, young workers, as part of their general training, should be given information relating to, inter alia, "the opportunities for general and vocational training", "the opportunities for promotion and the requirements to be fulfilled for the purpose" and "staff assessment systems". [Article 7(1)]

39. In France, Book IX: Continuing Vocational Training as Part of Life-long Education, of the Labour Code (Part I: Laws) as amended on 28 January 1981 states that "continuing vocational training shall form part of life-long education. Its purpose shall be to enable workers to adapt to changing techniques and conditions of work, to facilitate their training for promotion by affording them access to the various levels of culture and skill and to enhance their contribution to cultural, economic and social development" [L.900-1]. The types of training schemes falling within this provision include "adaptation schemes" to enable a worker to take up a first job or a new job more easily, "schemes for promotion", "schemes for acquiring, maintaining or improving knowledge" to enable workers to maintain their income and "to assume greater responsibilities in their working lives" [L.900-2].

40. The Works Constitution Act of 1972 of the Federal Republic of Germany also specifies that employers and works councils shall ensure that "workers are given an opportunity to participate in vocational training programmes inside and outside the establishment". [Article 96]

Regulations concerning visual display units (VDUs)

41. A prominent example of a particular technological development in respect of which several countries have issued specific regulations is the visual display unit (VDU).

42. In the Federal Republic of Germany, national regulations on VDU use were issued by the Industrial Injuries Institute on 1 January 1981. The Industrial Injuries Institute is a joint management-union body which has the right under the 1973 Work Safety Act to introduce regulations on specific aspects of occupational safety and health. The regulations deal mainly with design characteristics and standards for component parts (keyboard, chairs, illumination, work layout, etc.) of VDU work stations. These standards are determined by the Federal Standards Institute (Deutscher Institute für Normung - DIN). The regulations also contain a section on "Operation" of which certain provisions have implications for work organisation. Thus, the text specifies that in informing and instructing employees on the use of the VDU, "it is not sufficient" to give instructions only on the "manual and technical operation of the equipment, since the degree of possible strain depends largely on the employee motivation, knowledge of job-related work processes and the meaningful usage of the equipment provided" [Paragraph 6.2]. It further recommends that "the ergonomic aspects of the work should be taken into consideration as well as the ergonomic design of

display workplaces ..." [Paragraph 6.8]. Finally, it recommends that "from the ergonomic point of view, the relaxation achieved by several short work interruptions ... is very much higher than fewer and longer pauses" [Paragraph 6.8].

43. In Sweden, the National Board of Occupational Safety and Health issued Directive No. 136 on "Reading of Display Screens". These regulations also consist mainly of technical recommendations concerning equipment and workplace layout design and operation.

44. In Austria, the Ministry of Social Affairs has issued an ordinance concerning provision of rest pauses for workers using visual display terminals. The ordinance provides for additional rest periods when working with the VDU constitutes the essential part of the overall work performed. Moreover, where work on a VDU is performed for two or more hours without interruption, a break of ten minutes, to be credited as time worked, is to be included for each 50 minutes of continuous work.

45. In Japan, the Ministry of Labour notifications of 1964 and 1975 concerning the regulation of work-rest schedules limit the length of each work spell to 60 minutes followed by a break of 10-15 minutes.

B. Collective bargaining

46. In many countries, the scope of collective bargaining has been expanding to cover technological questions either through the inclusion in existing agreements of clauses dealing with technological change or through the conclusion of new technology agreements. New technology agreements tend to be more comprehensive in their approach to technological change than conventional types of agreements. This section contains examples, drawn both from conventional agreements and from new technology agreements, of provisions on procedural and substantive issues relating to the introduction of new technology and work organisation.

Objectives

47. The introductory clauses concerning objectives aim at establishing the common ground between management and workers. They usually include statements about the

benefits of new technology and some reference to conditions governing its introduction. However, agreements vary with respect to their explicit consideration of social effects of new technology on workers. Examples of such clauses are the following:

- "All parties recognise that the introduction of computer-based systems and equipment can be to their mutual advantage by improving efficiency" [Memorandum of Agreement between Cossor Electronics Ltd., Harlow, and APEX, ASTMS, AUEW-TASS and Engineering Staff Association, 7 July 1982 (United Kingdom)];

- "The parties declare their positive approach to the utilisation and development of new technology which may improve competitiveness, employment, the working environment and job satisfaction in undertakings" [Supplementary Agreement between the LO and the Danish Employers' Confederation (DA), January 1981 (Denmark)];

- "Standing on the common recognition that the progress of technology is indispensable to the continuation and development of a company ..., the Company and the Union shall always be mindful of the effects which the introduction of new technology could bring about on employees ..." [Agreement concerning the introduction of new technology between Nissan Motor Co., Ltd., and the All Nissan Motor Works' Unions, 1 March 1983 (Japan)];

- "... it is important that computer-based systems are evaluated, not only from technical and economic angles but also from social angles, so that all the aspects are taken into account in the development, introduction and use of such systems. This includes changes in organisation, employment, information routines, human relations ..." [General Agreement between LO and the Norwegian Employers' Federation (NAF), April 1975, revised in 1978 (Norway)].

Technological scope

48. Agreements tend to differ in their definition of the technology covered by the agreement. In some cases, the definition seems to imply any technological change. In others, technology is defined relatively broadly to cover a whole range of new technology. In this case, terms like "computer-based" or "micro-processor-controlled" systems are commonly used. Some agreements, however, cover only a

specific type or piece of equipment or a single process. Typical wordings of these definitions are given below:

- "In the context of this agreement, New Technology shall specifically relate to the introduction of new equipment and systems (including computer-based systems) which may change the nature in which tasks are performed" [Agreement on New Technology between Rolls-Royce Ltd., Leavesden and Hatfield, and APEX, 30 June 1981 (United Kingdom)];

- "For the purposes of this agreement, new technology shall be taken to mean new or modified computers or micro-processor-controlled equipment used in the Civil Service which have different staffing requirements from those of equipment already in use, or which result in changes in work processes, techniques or the allocation of work, which are of major significance to those directly affected by them" [Agreement on New Technology between National Whitley Council and the Council of Civil Service Unions, 22 March 1982 (United Kingdom)];

- "After consultations with the workers' representatives, the company intends to introduce into all the company stores visual display equipment belonging to the IBM 5280 system including all the peripherals that go with it (control unit with keyboard, diskette stacks, VDU, printer, telephone connection) ..." [Agreement between the Company Management and the Central Works' Council of F.W. Woolworth Co. GmbH, Frankfurt-on-Main, 7 December 1982 (Federal Republic of Germany)].

Advance notice and provision of information

49. These are usually the first steps in regulating new technology through collective bargaining. They are essential to the consultation and negotiation process. The timing, form and content of the information to be supplied are important to enable workers and their representatives to participate and to prepare for the changes to be implemented. The early disclosure and discussion of available information can also avoid costly investments which prove unsuitable or delays in introduction because of opposition from the workforce.

50. Advance notice from management is often required to give workers and their representatives an interim period or "lead time" to consult, negotiate or prepare for the forthcoming changes. Provisions concerning advance notice

can be specific by stating the required number of days of the period of notification. A 30-90 day notice is relatively common. In some cases, provisions are general and phrases like "advance warning", "informed as early as possible", etc., are used. Typical examples of such provisions are given below:

- "The Employer agrees that prior to the introduction of any technological change ... to give the Union at least ninety (90) days prior notice ..." [Article 17 - Technological Change of the Agreement between the Bank of Montreal, Windsor Main Office Branch, and the Union of Bank Employees (Ontario), 1 January 1983 (Canada)];

- "The company pledges to give to the local works' council advance warning of the installation of the IBM 5280 ..." [Agreement between the Company Management and Central Works' Council of F.W. Woolworth Co., GmbH, Frankfurt-on-Main, 7 December 1982 (Federal Republic of Germany)].

51. Some provisions specify disclosure at the planning stage and participation of workers in job design, as noted below:

- "The employer is obliged, already at the planning stage, to provide information on significant changes in the work duties, in the working place and in the working conditions ..." [Central Agreement on Information between Employers' Confederations and NBU-Affiliated Organisations (Finland)];

- "The employees shall, individually or in groups, be given proper information ... about conditions at the workplace that affect their own job ... The employees shall be given an opportunity to take part in designing their own job situation as well as in the work of change and development that affects their jobs" [Agreement on Efficiency and Participation SAF-LO/PTK (Sweden)].

52. Clauses relating to provision of information can also specify areas or subjects on which information should be provided.

- "It must be stated clearly in the notification and consultations what technical, organisational and/or staff changes are intended and how they are to be carried out. The documents must provide information on the following points in particular:

  - the objective and economic dimension of the system;

- the direct and indirect rationalisation measures which the system will entail and which will affect the workforce;

- the extent of the effects on the tasks and jobs established hitherto;

- the changes in working conditions (in the quality of work, the work process and job design);

- the utilisation of job-related equipment;

- the plants, departments and jobs which will be affected".

[Plant Agreement between the Management and General Works Council of Volkswagen Werk AG, 1 June 1980 (Federal Republic of Germany)]

53. In Italy, the national collective agreement for the metal trades specifies that information concerning production outlook, research and development policy, investment programmes, effects on employment and working conditions, professional training, decentralisation of production, etc., should be provided to trade union representatives.

54. In some cases, provisions also contain requirements concerning the form in which the information must be presented. Thus, the information provided must be in terms which are understandable to workers and/or their representatives who have a broad knowledge of the principles of computer-based systems but who do not have a technical background. An example of such a provision is given below:

- "The information will be given clearly and in a language easily understable to those without specialist knowledge in the area concerned" [General Agreement between the Norwegian Federation of Trade Unions (LO) and the Norwegian Employers' Confederation (NAF) (Norway)].

55. Some agreements also require that managers should provide relevant information on a continuing basis. For example:

- "Continuous information shall also be provided about any important amendments to plans which have been put forward" [Technology Agreement supplementing the Agreement on Co-operation between LO and DA, 1 March 1981 (Denmark)];

- "... the Corporation will make available sufficient information to the union ... to enable it to (i) monitor developments, changes in workflow, changes in working methods and the effects on jobs ..." [Agreement between the General Accident, Fire and Life Assurance Corporation Ltd., and its Subsidiary Companies, and APEX, 7 August 1980 (United Kingdom)].

Consultation and negotiation

56. These are procedures to be applied prior to the introduction of new technology. As previously mentioned, in some countries these are provided for in legislation.

57. Consultation implies that management must hold discussions with workers and their representatives (works councils or trade unions). In some cases, consultation is limited to providing information or advance notice about impending changes. Negotiation, on the other hand, implies that change will only be introduced after completion of a structured set of formal discussions at the appropriate levels of the enterprise. Moreover, in some cases where agreements provide for negotiation, deadlocks are subject to arbitration or referral to appropriate grievance and disputes machinery. The obligation to negotiate does not necessarily imply an obligation to reach agreement. In practice, however, it is becoming more difficult to draw a line between provision of information, consultation and negotiation proper.[3] This is particularly true because the introduction of new technology rarely involves a single isolated decision but triggers off a chain reaction with implications for workers at different levels. Management and workers, therefore, tend to engage in continuous information exchange, consultation and negotiation.

58. The following are examples of clauses on consultation and negotiation. In some cases, only consultation (which seems limited to "discussion") is provided for.

- "The company agrees to joint discussion regarding the introduction of new technological equipment once firm proposals are available ..." [Agreement between the International Harvester of Great Britain Ltd. and APEX, 4 February 1980 (United Kingdom)];

- "If requested to do so, the employer will meet with the Union to discuss the implementation of such changes before putting such changes into effect" [Retail Clerks and Giant Food Stores, 6 September 1980 (USA)];

- "The Company agrees that if and when a decision is reached by the management of the Company to install electronic data processing equipment, the Brotherhood will be informed of the decision. Any changes that affect the members of the Brotherhood will be negotiated" [Agreement between Niagara-Mohawk Corp. and IBEW (USA)].

59. In other cases, it is specified that new technology can only be introduced with union agreement. For example:

- "1. No changes in existing working and professional practices shall be introduced as a result of computerised photo-composition, or any other form of new technology, without prior agreement with the Chapel.

  2. The Management shall consult with the Chapel before any future technological change, and recognise the need for agreement where the working conditions of staff are altered".

  [Appendix 4 to the House Agreement between the National Union of Journalists Chapel and TRN Ltd. (United Kingdom)].

60. Some agreements also contain corollary status quo clauses which state that no new equipment or changes to existing equipment, procedures, etc., will be implemented until joint agreement has been reached. They also usually make reference to disputes procedures to resolve deadlocks in negotiation. Typical wordings of such clauses are the following:

- "Management will seek formal union agreement to operate each new system of equipment and existing systems. Management will not introduce such equipment prior to agreement being obtained with the union ..." [Agreement between International Computers Ltd. (PCM Kidsgrove) and APEX, 7 June 1979 (United Kingdom)];

- "All consultation/negotiations under this agreement shall be initially carried out departmentally by representatives of NALGO and Management. Any agreement reached will be endorsed by both sides of the Joint Consultative Committee before introduction takes place.

  Should the departmental negotiations fail to reach agreement, a status quo will apply and the matter shall be referred to Establishment Committee via the Joint Consultative Committee for their consideration.

If the matter cannot be resolved at this stage, it will be referred to the Joint Secretaries of the Provincial Council for conciliation. The status quo will remain in operation until a mutually satisfactory agreement is reached" [Agreement between the Sainthorpe Borough Council and NALGO (United Kingdom)].

61. With respect to consultation and negotiation procedures, it is important to note that in some countries legislation on co-determination or industrial democracy confers rights on workers and their representatives. This legislation has often been the basis for national agreements which in turn encourage consultation and negotiation at enterprise level. For example, in Sweden, the Central Agreement on Efficiency and Participation between SAF-LO/PTK (1982) was based on the Co-determination at Work Act. This central agreement stipulates that "participation and co-determination forms shall be adapted to local circumstances at the workplace" and that the local parties have joint responsibility for bringing about suitable participation and co-determination practices (Section 6, Article 1). In addition, "co-determination matters are dealt with by means of negotiation between the parties" (Section 7) who "should conclude an agreement upon the way in which co-determination shall be exercised" (Section 8, Article 1).

Access to outside expertise

62. Provisions on this point are based on the recognition that certain issues proposed for consultation or negotiation are of a highly technical nature. Since employers often use outside consultants or experts to formulate proposals or system designs, a similar right is also given to workers and their representatives in some agreements. These experts can be drawn from within the enterprise, the trade union organisation, universities, consulting firms, etc. In some cases, these experts are financed by the employer. Examples of clauses referring to access to outside expertise are:

- "Where the Association recognises a need for any particular technical expertise to aid their understanding of current issues, they may, by mutual agreement, co-opt additional representatives to take part in these meetings (divisional or local) in order to provide the necessary specialist knowledge in an advisory capacity" [Agreement between Imperial Chemical

Industries Ltd. (Mond Division) and ASTMS, 16 January 1981 (United Kingdom)];

- "The employees' delegates shall be enabled, in agreement with the management, to use the undertaking's own expertise to a reasonable extent (as regards time, priority). If necessary, the employees' delegates shall be able, in agreement with the management ... to consult outside experts ... The cost of such expert assistance will be borne by the undertaking unless some other course is agreed on" [Blanket Agreement on Technological Development and Computer Systems between NAF and LO, November 1981 (Norway)].

Training for workers' representatives

63. Some agreements specify that workers' representatives should be given opportunities to participate in training courses on new technology in order to carry out their duties. An example of such a provision is the following:

- "With a view to enhancing their effectiveness, Association representatives ... may be nominated to attend Trade Union or Trade Union appointed courses. The Society will not unreasonably withold permission to attend such courses" [Agreement between the Royal London Mutual Insurance Society Ltd. and ASTMS (United Kingdom)].

Job content, work organisation and job design

64. There is considerable variation in the coverage of job content, work organisation and job design issues in collective agreements. Some agreements merely make references to certain dimensions concerning the quality of jobs which should be considered in the introduction of new technology. Other agreements provide more detailed safeguards against possible negative effects. The issues commonly mentioned or covered are job satisfaction, job content, skills, workload, work measurement and performance monitoring, careers, and job grading and job evaluation schemes. Many of the agreements examined also specify that these areas will be the subject of negotiation and consultation. Examples of clauses concerning the abovementioned issues are given below.

65. Job satisfaction, job content and job enrichment. Some agreements specifically mention the need to consider job satisfaction and to avoid monotonous work. An example of this is the Agreement for the introduction of new technology between the Lucas Group of Companies and APEX/ACTSS in the United Kingdom, which states:

- "Every effort will be made to ensure that the content of jobs concerned with new technology avoids unnecessarily repetitive routine and emphasis will be placed on providing job satisfaction for those involved."

66. Some agreements specify these needs in greater detail. Examples of these agreements are the following:

- "The Tavistock Institute of Human Relations has listed the following as being relevant to creation of job satisfaction.

  Job content to be reasonably demanding in mental terms and provide some variety.

  Potential to learn and to go on learning the job.

  Some minimal areas of decision making that can be called one's own.

  Some minimal degree of social support and recognition in the workplace.

  Feeling that what is being done is significant and meaningful.

  Feeling that the job leads to some sort of desirable future.

  Both parties to this agreement accept that these principles will be incorporated, as far as possible, into the design of new working methods arising from the introduction of New Technology." [Agreement between the London Borough of Haringey and NALGO (United Kingdom)].

67. In Sweden, the General Local Agreement for Rationalisation and Administrative Development between the National Government Pay and Pensions Board (SPV) and the Statsjänstemannaförbundet (ST), a union of civil servants, contains extensive provisions concerning work organisation and job satisfaction. The relevant extracts from this agreement are given below:

- "Clause 1. Work organisation.

- Working routines and organisation are to be designed in such a way that employees will become thoroughly acquainted with SPV's activities and will have opportunities of good comradeship at work.

- The employees are to be able, both individually and on a group basis, to influence the conduct of work.

- Increased fragmentation of jobs is to be avoided.

- During rationalisation, organisational development shall precede or progress parallel to the development of technical aids.

- SPV shall aim for extensive delegation of decision-making powers and for the shaping of work organisation in such a way as to reduce and minimise the number of decision-making levels.

Clause 2. Job satisfaction.

- When new technology is introduced, the professional knowledge of the employees must be maintained at such a level that the requirements of legal security and good service can be accommodated.

- Computerisation must not be taken to such lengths as to detract from interesting and developmental duties."

[Agreement concerning rationalisation and administrative development between SPV and ST, 19 October 1981 (Sweden)]

68. The Agreement on Efficiency and Participation between SAF-LO/PTK in Sweden states that:

- "In the event of technical change, a sound job content shall be the goal, together with opportunities for the employees to increase their skills and accept responsibility for their work. The knowledge of the employees should be stimulated together with their ability to co-operate with and have contact with their colleagues."

69. In the Federal Republic of Germany, as long ago as 1973, the North Württemberg-Baden agreement in the metalworking industries considered issues relating to the

design of assembly jobs, job enrichment, group production and work cycles.[4] Specifically, it states that the design of assembly-line jobs and similar work consisting of repetitive operations performed at regular intervals shall mainly aim at job enlargement and job enrichment in order to alleviate the adverse human effects of monotony. This obligation is particularly binding in cases where the job time or the work cycle is less than 1.5 minutes. The agreement also states that the pace of work at the flow line should be based on the time required for the longest operation. Furthermore, the agreement stipulates that jobs, working methods and the working environment will in the future be designed with people in mind.

70. Again, in the Federal Republic of Germany, the works council agreement concerning the introduction of VDUs within Volkfürsorge (Insurance Co-operative Company) states that whenever possible VDU operators should carry out other tasks so that they perform "composite jobs".

71. Some agreements also include clauses concerning downgrading. An example of a general provision is the following:

- "The company shall not effect any downgrading of positions or any reduction in wages and working conditions of affected members for reasons of the introduction of new technology" [Agreement concerning the introduction of new technology between the Nissan Motor Co., Ltd., and the All Nissan Motor Works Union, 1 March 1983 (Japan)].

72. On the other hand, there seems to be a distinction between a guarantee that no employee will be downgraded and a guarantee that no job will be downgraded. Most agreements refer to the former. This usually means that an employee can be re-deployed to a lower position but guaranteed the same salary grades and increments. For example, the Agreement on Downgrading Arrangements between the Imperial Tobacco Company and GMWU (United Kingdom) sets out in detail how employees are to be protected from loss of contractual earnings when they are downgraded due to changes in work methods and production requirements. The period of retention of contractual earnings depends on the period of service in the company, age and "regular set overtime". A sliding scale is set out in the agreement. Typical wording of such a provision is the following:

- "An employee re-deployed to a lower grade as a result of the introduction of new technological equipment is guaranteed to maintain his/her grade and incremental status as at the time of transfer. In addition, all salary awards and any other general increases negotiated will be paid without off-setting ..." [Agreement between International Harvester Company of Great Britain Ltd. and APEX (United Kingdom)].

73. Aside from the general reference in the Nissan agreement mentioned above, none of the agreements examined for this report contained guarantees that jobs will not be downgraded. However, as previously mentioned, several agreements included clauses aimed at minimising monotony and facilitating job satisfaction, job enrichment, etc. Moreover, many of these agreements specify that such changes in jobs would be considered in the consultation or negotiation process.

74. Workload. Some clauses in agreements have direct implications for workload. An example of this is:

- "Staff operating the new systems will not be given additional tasks or added responsibility without prior consultation" [Agreement between Cossor Electronics Ltd., Harlow, and ASTMS, APEX, AUEW/TASS and Engineering Staff Association, 7 July 1982 (United Kingdom)].

In some cases, provisions concerning training have indirect implications for workload. For example, the clause in the agreement concerning employment at video terminals between the German Postal Workers Union (DPG) and the Federal Minister of Posts and Telecommunications specifies that "an employee shall be given adequate time and opportunity to familiarise himself with the apparatus".

75. Job evaluation. Several agreements provide that job evaluation schemes and/or new jobs should be examined in order to take into account changes in skill requirements, qualifications, job demands, etc., due to the introduction of new technology. Some agreements provide that new jobs will be evaluated based on the existing job evaluation scheme. An example of such a clause is the following:

- "Technological changes could affect the skill requirements and occupational categories of many staff. The established evaluation system will be used to measure the relative values of jobs when changes take place ..." [Agreement between General Accident, Fire and Life Assurance Co., and ASTMS, 7 August 1980 (United Kingdom)].

On the other hand, some agreements state that the new jobs and the existing job evaluation schemes should be re-examined in the light of technological changes. An example of such a provision is the following:

- "Full analysis and consultation will take place in respect of the effects on the existing job evaluation scheme and on jobs evaluated under the scheme caused by the introduction of new systems" [Agreement between Cossor Electronics Ltd., Harlow, and APEX, ASTMS, AUEW/TASS and Engineering Staff Association, 7 May 1982 (United Kingdom)].

76. Monitoring of work performance. Many of the agreements examined either prohibited the use of machines to monitor work performance or allowed monitoring only with union agreement. An example of the first kind is the Agreement between the Company Management and the Central Works Council of F.W. Woolworth Co., GmbH, Frankfurt-on-Main (Federal Republic of Germany).

- "Data collected by IBM 5280 system shall not be used to control or evaluate the performance of personnel working on the equipment".

An example of the second is provided by the Agreement between National Whitley Council and the Council of Civil Service Unions in the United Kingdom:

- "Work measurement is currently used to monitor and test the use of equipment and to provide information for the proper management of the work. Management will not use new technology in new ways to provide data for use in the assessment of the performance of an individual without first consulting and seeking the agreement of the appropriate trade union side ...".

Training or retraining for workers

77. Many of the agreements examined contain clauses concerning training or retraining for workers. However, there is considerable variation in the coverage and specifications of such issues as who gets the training, the scope and content of training programmes, the length of training, who pays for the training, etc.

78. Some agreements contain only a general statement that training will be given to those who will be involved in the change. However, some agreements are more specific as to who should be trained. An example of such a clause is the following:

- "The company is committed to all appropriate forms of training. It is agreed therefore that:

(a) Training programmes will be established for those employees who will use the new equipment/systems.

(b) Employees redeployed as a result of the introduction of new technology will receive appropriate retraining to equip them with the necessary skills.

(c) For those employees in jobs which although are not directly or significantly affected by the introduction of new equipment but who need an understanding of the system, new technology appreciation courses will be progressively arranged. The selection of such employees will be discussed and agreed between the parties to the agreement."

[Agreement between Rolls-Royce Ltd. (Leavesden and Hatfield) and APEX, 30 June 1981 (United Kingdom)].

79. In many cases, training is restricted to the users of the equipment. The following clause is typical:

- "All employees who will or could be required to use an equipment as part of their normal duties will receive full relevant training" [Agreement between GBS Turbine Generators Ltd. (Willans Works, Rugby) and ASTMS, 1 November 1982 (United Kingdom)].

80. In some agreements, particularly in the United States, a common condition is that the employee must be capable or qualified to be retrained. The following clauses are typical:

- "In the event training programmes are required to develop skills that may be required to fulfil this paragraph, the company agrees to train such displaced or reassigned employees who are capable" [Agreement between IAM and Jor Manufacturing].

- "Whenever new equipment or devices are introduced, it is the company's intention to train current employees who have (i) kept abreast of technological changes, and (ii) possess the capabilities of performing these job functions" [Agreement between IATSE and WJZ-TV].

81. In some agreements, the scope and content of training programmes are specified. The following clauses are representative:

- "... the course content, which has yet to be formalised, will basically consist of the following subjects: -

  (1) General appreciation of computer and associated terminology.

  (2) Appreciation of VDU and associated systems.

  (3) Health and safety implications.

  (4) Mechanics of operation.

  (5) Familiarisation of other staff."

  [Agreement between NEI Parsons Ltd. and APEX, 24 May 1979 (United Kingdom)].

- "Before being employed at a video terminal post for the first time, a worker must in good time be given comprehensive induction into working methods and the handling of the apparatus. In particular, he must be acquainted with the methods of adjusting and handling it dictated by the principles of ergonomics. Induction may be supplemented by further training ... where this is necessary on account of the special features of performance of the work to be done with a video terminal" [Agreement concerning employment at video terminals between the German Postal Workers Union (DPG) and the Federal Minister of Posts and Telecommunications (Federal Republic of Germany)].

82. Some agreements indicate the duration of the training to be provided. This provision can be general and open-ended such as the Agreement between Bonds (NZ) Hosiery Division and the Wellington, Taranaki and Marlborough Clerical, Administrative and Related Workers Industrial Union of Works in New Zealand.

- "All workers required to use any aspect of the new technology will be provided with sufficient training to enable them to perform competently to the satisfaction of both the employee and management."

In some cases, the length of the training period is specified explicitly.

- "The Employer will provide a training period of up to fifteen (15) days to an employee who is thereby displaced provided that she has the skill and ability to perform the full requirements of the job within such period" [Agreement between the Bank of Montreal,

Windsor Main Office Branch, and the Union of Bank Employees (Ontario), 1 January 1983 (Canada)].

With respect to who will pay the cost of training, most agreements specify that management will pay.

II. Legislation and practices in selected countries with planned economies

83. In the Central and Eastern European countries with planned economies, legislation and collective agreements also deal with issues arising from the introduction of new technology. But they are often supplemented by plans, models, standards and institutionalised practices that have a significant influence on working conditions and work organisation. An important common feature is the extensive role assigned to trade unions in decisions affecting work organisation.

84. A brief description is given below of the relevant legislation and other normative texts or institutionalised practices in a number of these countries.

85. In the USSR, the objective of using new technology to improve work organisation and working conditions is specified in the Constitution:

- "The State concerns itself with improving working conditions, safety and labour protection and the scientific organisation of work, and with reducing and ultimately eliminating all arduous physical labour through comprehensive mechanisation and automation of production processes in all branches of the economy." [Article 21]

86. Legislative provisions in the countries under consideration define the rights and obligations of management and the role of trade unions with respect to the organisation of work.

87. In the German Democratic Republic, for example, the Labour Code of June 1977 (Chapter IV: Organisation of Work and Socialist Labour Discipline) specifies that the undertaking should provide conditions of work that "enable a worker to achieve a high standard of performance, encourage a conscious and creative attitude to work, increase job satisfaction and contribute to the development of a socialist personality ...". Emphasis should also be placed on the "creative elements of work" in the design of

jobs and on limiting workplaces "involving physically arduous jobs or jobs involving unilateral forms of strain". Moreover, regulation of working conditions should be based on the latest scientific knowledge. The Labour Code also gives trade unions extensive rights such as participation in the preparation and application of plans for the national economy, rationalisation measures, output standards and performance measurement in the undertaking.

88. In Czechoslovakia, the Labour Code gives the trade union committee in the undertaking the power of co-decision with management in certain fields such as the amendment or publication of work regulations and improvement of the working environment and the introduction or amendment of pay systems. Under this power of co-decision, the trade union committee may insist that a subject shall be submitted for its prior consent.

89. In the USSR, the trade union committee has a right of co-decision on work rules, output norms and the manning of work teams. It also has extensive powers of supervision to ensure that the provisions of legislation and collective agreements are observed.

90. Similarly, in Hungary, the Act concerning the management of State undertakings, which came into force on 1 January 1978, gives workers' representatives co-decision rights on questions concerning conditions of work (e.g. output norms, hours of work, pay).

91. In these countries, the undertaking usually has the responsibility for training and retraining of workers. However, the management and trade union committees are called upon to collaborate in preparing, running and supervising the various training schemes. For example, the Labour Code of the German Democratic Republic (Chapter VIII: Training and Further Training) specifies that the undertaking shall ensure that the "training and retraining required in connection with rationalisation measures or structural changes shall be planned and carried out in such a way that workers have the necessary skills when taking over some new or different forms of activity". This training and further training should be based on the "latest scientific knowledge and the most progressive experience acquired in the course of practical work". Moreover, the development, running and supervision of these training programmes should be done jointly by the management and trade unions committee of the undertaking. The cost of training is borne by the undertaking.

92. Legislation encourages the conclusion of collective agreements and specifies the procedures to be followed. It also often indicates certain areas to be covered by agreements. In the USSR, for example, the Regulations of 26 August 1977 governing the procedure to be followed in the conclusion of collective agreements provide that a collective agreement shall contain, among others, the following main sub-divisions: "introduction of achievements in the field of science, technology, pioneer experience and the scientific organisation of work", "remuneration and the setting of output standards", and the "improvement of workers' skills, economic knowledge and general lack of education". In addition, it should stipulate the mutual obligation of management and workers concerning, inter alia, "the improvement of labour productivity, the reduction of labour intensive and manual operations, ... rationalisation schemes, pioneer experience and the scientific organisation of production, work and management, and the mechanisation and automation of basic and subsidiary processes", and "the timely introduction and exploitation of production capacity and the reconstruction of existing production lines". Accordingly, under the collective agreement of the Volga Automobile Factory at Togliatti, for example, the management and the trade union committee of the undertaking are jointly responsible for introducing and developing standard times for technical processes and for determining the organisation and technical conditions which will permit all workers to meet those standards.

93. Another instrument of growing importance in these countries is social development planning.[5] Social development plans are now an integral part of comprehensive social-economic-technical plans of the undertaking. They are linked with the economic targets established in its five year plan and usually foreshadow the probable content of future collective agreements in the undertaking.

94. The content of social development plans varies according to the country and the specific characteristics of a given enterprise. However, the plans usually cover changes in the social structure of the enterprise or undertaking, major aspects of manpower policy (selection, training, promotion), problems and improvements concerning working conditions, work content and welfare facilities, and possibilities for increasing workers' participation.

95. Social development planning presents an opportunity to influence directly the socio-technical design of new systems or the re-design of existing ones. The planning process, which includes conducting sociological studies, attitude surveys, etc., at the undertaking, facilitates

consideration of the likely negative social effects of new technology at an early stage.

96. Most of the measures covered in these plans are operationalised and/or monitored by the social planning service of the undertaking. The size of this service or unit varies according to the size of the enterprise. In a small enterprise, there may be only two people while a big enterprise like the Volga Automobile Company in the USSR has a whole department. However, when long-term social development plans are to be formulated, this service unit becomes one of the members of a joint committee which also consists of representatives from management, trade unions and the party. This joint committee is assisted by other functional departments and expert groups (e.g. sociologists). In Czechoslovakia, the expert group is composed of "social development specialists" who are specifically trained in the social aspects of work.

97. A major role in the improvement of work organisation is also played by work organisation models and standards which have been developed and widely introduced in a number of Eastern European countries. For example, since their introduction in the early 1960s, there are now more than 3,500 models of work organisation covering six million work stations and involving more than 30 per cent of the labour force in the USSR.[6]

98. Work organisation models and standards are based on scientific research and economic objectives. They draw on the methods and results of research in various fields of knowledge and convert them into guidelines and methodological principles for designing or redesigning work processes, jobs and enterprises. They can be adopted directly or slightly adjusted to suit the conditions in a given undertaking, workshop or workplace. Their use minimises duplication of design work and enables the undertaking to take advantage of the achievements of science and technology.

99. The first models of work organisation were developed for individual work posts and certain work processes, but developments in technology and the emphasis on collective forms of work organisation led to the development of more comprehensive models such as complete production sections or shops. The basic idea is to arrive at a uniform model for standard jobs, work stations or production departments which incorporates the best possible solutions for all aspects - technical, economic, physiological, social - of work organisation.

100. The content of the work organisation models varies according to the type of production process, when the model was developed and which design or research institute was involved. None the less, most models are likely to cover the following:[7]

(a) definition of the scope of each scheme (description of the machines and type of work to be performed to ensure proper selection by the user);

(b) description of equipment, tools and technical documentation attached to the work post;

(c) workplace arrangements or layout;

(d) specifications for communication, supply and dispatching arrangements;

(e) sequence of operations, financial and performance indices, quality standards;

(f) workers' qualifications and responsibilities;

(g) working environment, ergonomics and occupational safety and health considerations;

(h) cost benefit methodology.

101. Work organisation standards or norms are developed on the basis not only of technical efficiency (in terms of working time and equipment) but also of economic, physiological, psychological and sociological factors. They concern (a) the design of equipment, technological processes and undertakings; (b) work organisation; (c) working conditions; and (d) labour expenditure.[8] Observance of these standards and norms is compulsory.

102. The models and standards are developed by labour research and design institutes attached to the different branches of industry and by centres of scientific work organisation.

103. Finally, certain practices - backed up by legislation and the principles of the industrial relations systems - are aimed at promoting both workers' participation in the process of technological change and improvements in work organisation and working conditions.

104. In line with their effort to promote collective forms of work organisation and worker participation, many

Socialist countries have introduced "work brigades".[9] The basic organisational principle of the brigade is the manufacture of a product or a complete component of a product by a group of workers whose tasks generally form part of the same technological process. Within the brigade there is always the possibility of switching jobs. This possibility is intended to facilitate improvement in work content (job rotation, job enrichment, etc.), worker versatility and production flexibility.

105. The brigade system of work organisation also expands workers' participation in decision-making at the level of the brigade, the workshop and the undertaking. The members of a brigade elect a "brigade council" which acts as a link between the team leader ("brigadier") and the work brigade. The council examines draft production plans; reports on the progress of the brigade with respect to the production schedule; sets and evaluates the contribution of each member in obtaining production goals; selects brigade members for promotion; decides on the brigades' composition; decides and distributes bonuses and initiates proposals concerning improvement of working conditions, revision of production quotas, etc. Some of these functions were formerly the exclusive responsibility of foremen and heads of sections. The council of brigade leaders ("council of brigadiers"), which co-ordinates inter-brigade linkages at plant level, works closely with top management and has considerable bargaining power.

Notes:

[1] Those texts that have been published in the ILO Legislative Series are listed at the end of this paper.

[2] For a detailed analysis of Section 12, see: B. Gustavsen: "A legislative approach to job reform in Norway", in International Labour Review (Geneva, ILO), May-June 1977, pp. 263-276; and B. Gustavsen: "Improving the work environment: a choice of strategy", in International Labour Review, May-June 1980, pp. 271-286.

[3] G.C. Pevare: Final Report on the Round Table Meetings on Information, Consultation and Negotiation Procedures for the Introduction of New Technologies, report prepared for the Commission of the European Communities, January 1983, and presented to the ETUC Conference on New Technologies and Working Conditions, Paris, 9-11 May 1983, p. 28.

[4] W.H. Staehle: "Federal Republic of Germany", in ILO: New forms of work organisation, Vol. 1 (Geneva, 1979), pp. 79-106; European Trade Union Institute (ETUI): Redesigning jobs: Western European Experiences (Brussels), 1981, p. 95.

[5] A. Prigozhin and A. Louzine: "Automation in USSR industry (National Background)", in F. Butera and J.E. Thurman (eds): Automation and work design (North-Holland, Amsterdam, forthcoming), pp. 21-24.

[6] A. Louzine: Work organisation models in USSR, CONDI/T/1980/4 (Geneva, ILO, February 1980).

[7] A. Prigozhin and A. Louzine, op. cit.; monograph on the German Democratic Republic by H. Hanspach and A. Schäfer in ILO: New forms of work organisation (Geneva, 1979), p. 17; A. Louzine, op. cit., p. 5.

[8] See the monograph on the USSR by A.S. Dovba, et al., in ILO: New forms of work organisation, Vol. 2 (Geneva, 1979), p. 89.

[9] ILO: Policies and practices for the improvement of working conditions and working environment in Europe, Report III, Third European Regional Conference, Geneva, October 1979, pp. 41-43. See also A.S. Dovba, et al., op. cit., pp. 91-99.

References

The following legislative texts have been published in the ILO Legislative Series (LS):

Denmark

Act respecting the working environment [LS 1975 - Den. 1]

France

Labour Code (Part I: Laws) as amended on 28 January 1981 [LS 1981 - Fr. 1]

German Democratic Republic

Labour Code of the German Democratic Republic [LS 1977 - Ger.D.R. 1]

Federal Republic of Germany

Works Constitution Act of 1972 [LS 1972 - Ger.F.R. 1]

Netherlands

Works Councils Act of 1979 [LS 1979 - Neth. 1]
Working Environment Act of 1980 [LS 1980 - Neth. 4]

Norway

Act respecting workers' protection and the working environment (Work Environment Act of 1977) [LS 1977 - Nor. 1]

Sweden

Act respecting co-determination at work of 1976 [LS 1976 - Swe. 1]
Working Environment Act of 1977 [LS 1977 - Swe. 4]

Union of Soviet Socialist Republics

Regulations governing the procedures to be followed in the conclusion of collective agreements [LS 1977 - USSR 1]

United Kingdom

Health and Safety at Work Act of 1974 [LS 1974 - UK 2]
Employment Protection Act of 1975 [LS 1975 - UK 2]

AUTOMATION AND WORK ORGANISATION: POLICIES AND PRACTICES IN MARKET ECONOMY COUNTRIES

B. Gustavsen[1]

I. Work organisation and job content

A. Reasons for recent interest in improving work organisation and job content

1. We can identify three main reasons for an interest in issues of work organisation:

- Democracy at work
- Productivity
- Work environment (originally: workplace, health and safety).

The relationship between workplace democracy and the way work is organised is well recognised. The main argument is that a high degree of division of labour, "Taylorism", drains away the resources of the workers through destroying their competence, decision-making ability and their social contacts. Workers with little autonomy and competence, and no collective structures to bind them together, are generally unable to have an active, constructive relationship to their own situation and problems and are consequently unable to fully participate in a democratic system (see for instance, Pateman, 1970; Emery and Thorsrud, 1976).

2. As far as productivity is concerned, the main argument against highly fragmented work is not only because it is problematic from the point of view of participatory democracy, it also poses problems from an efficiency point of view. The fragmentation of a larger task such as building a ship or running an oil refinery into the very small task pieces prescribed in Taylorism, is not consistent with the nature of most tasks confronted in working life which demand that fragmentation stops long before the very small elements of Taylorism are reached. Otherwise, task interdependencies would continuously disturb the production process and create problems for the operators (see, for instance, Emery, 1978, for a review and reprint of some of the main contributions to this argument).

3. While the importance of organisation for work democracy and productivity is widely recognised, the

importance for workplace health and safety has not always received as much attention. Workplace health and safety today constitute very complex areas of understanding and improvement (Gustavsen and Hunnius, 1981). Whatever public and professional apparatus exists to deal with these problems, such as the public labour inspection or the occupational health services at the enterprise level, they generally fall far short of the resources and experiences needed to grapple successfully with broad and ambitious reforms, such as were aimed at by many countries in the 70s. Public and the professional resources can deal with some of the issues, but far from all. As a guiding principle, initiatives and remedies must emanate from management and workers at the workplace itself. There is a very strong need for worker participation to define problems as well as to suggest solutions. Hence the issue of worker participation becomes highly important though there are various conditions that need to be fulfilled for such participation to be effective (Gustavsen and Hunnius, 1981; Gustavsen, 1983 a). Among these, organisation of work has emerged as being of critical importance.

4. These three "reform concepts" imply, to some extent, different considerations and solutions. There are, however, also common themes in the sense that certain patterns of organisation are positively related to all three. However, it is necessary to make a distinction. Development of new forms of organisation generally demand worker-management collaboration which must be based on a recognition of where interests coincide and where they do not. This can most easily be achieved by making it clear that organisation of work can be changed for different reasons, and what these reasons are. The use of vague concepts such as "quality of working life" often means that this point is lost and can result in workers and managers disagreeing on "what the changes are about" (see, for instance, Wells, 1983).

B. Dimensions of work organisation and job content: criteria for improvement from a worker perspective

5. In order to examine the relevant features of work organisation from the workers' perspective, a large number of concepts have been introduced. In the light of the last ten years of research and practical experience, however, many of these can be simplified and concentrated. Listed below are the dimensions that emerge as the most significant to which others can be related as sub-issues.

6. The major dimension is control (over the work situation), self-management, or autonomy. In turn, this can be sub-divided into:

- Skills, or competence
- Autonomy, discretion, or the possibility to make decisions
- Communications.

Skills or competence refer to knowledge, insight and abilities, while autonomy, or the possibility to make decisions, refers to the utilisation of that competence. There is an interdependence between competence and autonomy: competence which is not used withers away, while decision-making is a socialisation process (Karasek, 1981) which enables people to absorb more and more information and develop new abilities. The addition of communications underlines that, in the workplace, skills and decision-making often have to be collective (Gustavsen, 1980 a). These dimensions can, in turn, be broken down into a large number of issues, a point which will not be further pursued here.

C. Impact of poor work organisation

7. Three main trends of thought, on the impact of poor work organisation, can be identified. One school of thought has particularly focused on psychological consequences. The chief concept has been that of dis-satisfaction, which is linked to a great number of other concepts (see, for instance, Seashore and Taber, 1976). In the formative phase of the human relations school, it was thought that job satisfaction had a strong bearing on productivity. Later investigation has brought to light some instances of correlations between job satisfaction and productivity, but these correlations are generally weak and, often, such a correlation cannot be found at all.

8. Another related school is more sociologically oriented. It emphasises the anti-socialisation effect of poor work organisation, such as lack of ability to participate in democratic processes. Passivity, which describes the main focus of this school of thought, works contrary to democratic values as well as those of productivity. Passive workers generally show little interest in either participation or the production process (Pateman, 1970; Emery and Thorsrud, 1976; practically all efforts emerging from the socio-technical school reflect this consideration, for a review, see Jenkins, 1983).

9. Thirdly, work organisation can also be related to the concept of ill health. This approach is related to the other two. For example, they all acknowledge that dissatisfaction - if it is strong and long-lived - can eventually generate psychosomatic processes which result in damage to health. The chief factor in linking work organisation to health has been the concept of stress. Different schools of thought also exist in stress research. There is a tradition which is relatively closely linked to medical research, where emphasis is on the physiological processes generated by stress, and where stress is defined in relation to such processes. Another tradition is more closely related to psychological research and overlaps strongly with the satisfaction/dissatisfaction school of thought.

10. Some researchers have attempted to integrate the various approaches and perspectives and relate them to the issue of workplace control (Gardell, 1976; Frankenhaüser and Gardell, 1976). Generally, the following model is the point of departure:

Work demands and other pressures —— Control —— Impact on the workers

11. Empirical studies have demonstrated that control is a critical intervening dimension between a number of pressures (or hazards) and a broad set of impacts, ranging from social via psychological to physiological. Generally, heavy work demands and a low degree of control constitute the most critical situation from a health perspective (Karasek, 1981).

12. Even though it seems reasonable to integrate the various approaches and perspectives, there is a need to keep in mind that some basic differences exist between the approaches in their view of people. Focusing on, for instance, satisfaction/dissatisfaction, implies seeing people as things or objects having internal states of a positive or negative character, induced by environmental factors in combination with general human "needs" (such as those indicated by Maslow, 1943, 1954). Focusing on control implies seeing man as an active, creative entity, that is, as a subject. This is a fundamentally different emphasis, even though actual research practices based on it do not completely neglect "states in people". However, research related to such "states" is subordinate to the dialogue between the researchers and those to whom the research pertains (Gustavsen, 1983 b).

D. Criteria of economy and efficiency

13. To the extent that one wants to find solutions to problems of work organisation on the basis of a merger between participation, work environment and productivity, or even on the basis of the productivity issue alone, it is necessary to consider what content concepts such as productivity, economy and efficiency should be given. There has been a development within this area. Traditionally, within the Taylorism framework, productivity has been defined with a strong emphasis on quantity, that is, in terms of units produced per amount of time. Such concepts of productivity lie behind most of the traditional performance-premium wage systems. Of course, there has always been a need to apply other measures of productivity as well, but these dimensions, such as the ability of the enterprise to innovate, have generally not been considered relevant to the way work is organised in ongoing production. Over the last decade a successive shift in emphasis has taken place in such concepts as productivity or efficiency. Some examples of new areas of emphasis are:

- Product quality
- Throughput time
- Ability to avoid/deal with disruption and production breakdown
- Ability to handle different product versions
- Minimising the time required for re-balancing and running-in when production volume or models are changed
- Ability to deal with variations in manning due to absences, holidays, etc., without replacements.

The so-called "new factories" programme in Sweden stresses such productivity considerations (Lindholm, 1979; Aguren and Edgren, 1979).

14. The increased emphasis on product quality, particularly in industrial mass production where quality has not traditionally been a chief consideration, has been very dramatic over recent years as competition has increased. This development is most clearly seen in the car industry where, some years ago, most people shrugged their shoulders at the efforts made by Volvo in Sweden to get away from assembly line production. Today, this industry seems to be scrambling around in attempts to create alternative production concepts: some try increased cycle time, group assembly, product workshops, etc.; others put their faith in automation, while some pursue both lines. Assembly in parallel groups, for instance, also has its costs, in terms of greater demand for space, parallel sets of tools, buffer stores, and more expensive

transport systems (Lindholm, 1979), but when product quality is an important issue, these costs are worth bearing. The emphasis on quality under increasing technological complexity also makes it continuously more difficult to separate the production of given products from the design of new ones. Instead, integration must be sought, for instance in the form of utilising CAD/CAM systems in ways which result in the active participation of the production workers in the development of new products.

II. Implications of automation for work organisation and work content

15. Automation can be seen as a sub-issue within the broader concept of technologically-initiated development. Technological development has been a topic of major interest for a number of years. Numerous reports, studies, etc., have been conducted to assess the impact of technological changes, skill requirements, employment, productivity strategies, etc. In spite of the many efforts to clarify the impact of technological development, we are still far from such a clarification. We seem to confront a paradox: on the one hand, technology penetrates more and more spheres and becomes of increasing importance; on the other hand, it seems highly difficult to pinpoint the effects of this development. What might be the reasons for this paradox?

A. Some hypotheses about technological development

16. In the first phase of industrialisation, technology consisted largely of separate tools and machines. When these were brought together in factories, it was as much for economic and legal reasons as for reasons of production engineering (Gordon et al., 1982). A second phase gained momentum around the turn of the last century and was characterised by machines and other production equipment being brought together into more comprehensive production systems. This transformation was accompanied by the development of certain principles and theories about an optimal production system (e.g. Taylor, 1913). These theories were eventually used almost universally. One can talk about an <u>implosion</u> in the sense that production systems were drawn towards certain general characteristics. In this period, the assembly line was the most widely described organisational principle. The continuous process is

another such general concept. Automation, defined as a gradual take-over by machine of generalised human capabilities (see, for instance, Bright, 1958; Amber and Amber, 1962) was a third principle. In this "second phase" of industrial development, the possibility of talking about technological development as a definable process with a set of specifiable consequences still existed. Does this phase still prevail? On the whole, it clearly does. However, elements of a new phase seem to be emerging. This third phase is still of limited importance, but its importance will grow. This new phase can be characterised as an explosion: development is "thrown" in a number of different directions and no longer contracts around a few general principles. In spite of the presence of a number of components which resemble each other, such as computers, it looks as if we are now in a phase where technological development takes place along a number of different lines with each line having specific characteristics: communication technology, transport technology, technology for small series mechanical production, etc. In addition, there are two related factors. Firstly, technology is becoming more and more flexible because technological components can be used in a number of different ways. Practically all computer technology is of this flexible kind: its use is not specified until operational procedures are decided on. Secondly, there is an increasing choice between different technological solutions to any given problem or function. An example is technology for internal transport in factories where different options exist. For instance, the choice between an assembly line and a trolley system has different organisational consequences. An organisation which believes in Taylorism will probably choose an assembly line while an organisation which does not believe in Taylorism may choose trolleys in order to bring about the possibility of group assembly. This brings us to the major point that the character of technology is organisationally indeterminate. The impact of technological development on such dimensions as qualifications, productivity strategies, work environment, etc., are to an increasing extent mediated by the patterns of organisation, beliefs, or criteria, that guide enterprise behaviour. Automation can be defined as a process of re-organisation that relies particularly heavily on the use of technology.

17. If technology, in spite of its obvious importance, becomes more flexible, it explains why it is so difficult to pinpoint the specific consequences of technological development. This point has emerged in several ways in recent years. For instance, German research on development of qualifications demonstrates that there is no single trend

to be found (Pröjektgruppe Automation und Qualifikation, 1978). Does this mean that any pattern of organisation whatsoever can be super-imposed upon technological development? Is the choice between organisational patterns a free choice?

18. There is one obvious constraint. If specific technologies become more and more rapidly outdated, the enterprise must be able to continuously introduce new technologies. This has certain consequences for decision-making capacity and for the ability to adapt to changes, etc. Technology offers a fairly open field of choices, but other factors enter the picture, e.g. markets, local and global economic conditions, etc. In recent years we have, for instance, seen a movement towards stronger competition on the world markets. Such factors indicate very clearly that some lines of development are better than others. To this must be added the conditions under which human beings are able to perform at their best, i.e. when work is associated with a reasonable degree of autonomy, competence and communication. And here we arrive at a very important point: an exploding technological development increases the number of options. Furthermore, superior options will tend to be chosen. For instance, as more and more alternatives to assembly line production emerge, some enterprises will choose the alternatives which create new man-maching relationships, such as group assembly. While it is possible for one company to use new technologies under "phase two" type thinking, this becomes difficult if some of its major competitors create new socio-technical patterns. When new technological possibilities emerge and are used to generate new and more humane patterns of work and organisation, then a new moving force has entered the scene, which cannot be neglected. The original technology of Ford will, for instance, probably have disappeared from the European car industry by the end of the 80s: those who keep on with it will not survive. Such a moving force is, however, not constituted by technology alone, but by technology within the context of an alternative organisational thinking. We will proceed by looking into this issue of organisational alternatives.

B. The meaning of alternative patterns of organisation

19. It has long been recognised that there are different options, or possiblities, of how to organise and the various alternatives have often been compared, e.g. Taylorism versus socio-technical thinking; mechanistic organisation versus organic; centralised versus decentralised, etc. The question is if the various pairs of alternatives that can be listed really oppose each other on the

same dimensions or if there is a need to distinguish between different levels of organisation. The latter approach would make a distinction between concrete organisation, i.e. the actual patterns exhibited by an enterprise, and meta-organisation, i.e. the ability of the enterprise to create one or more other patterns of organisation. If we consider two enterprises both having the same pattern of organisation, there may still be a major difference between them in that the one may be able, at reasonably short notice, to change its given pattern for a new one, while the other may be unable to. While both enterprises may start with a given form of organisation, the meta-organisation is different in that one is rigid while the other is innovative and change-oriented. Even the most dynamic enterprises can have "bureaucratic" characteristics. The challenge for the enterprise lies in its ability to break out of this pattern if circumstances so demand. Concrete and potential patterns of organisation are, of course, not independent of each other. Concrete patterns can easily influence meta-organisation.

20. The new forms of organisation discussed here are forms of meta-organisation which have the ability to create a broad range of concrete organisational forms. If the explosion hypothesis concerning technological development is accepted, the enterprises most ideally suited to benefit from the situation are those that can:

(i) Develop and maintain a number of different concrete patterns simultaneously, in order to be able to give a number of different functions an optimal organisational expression.

(ii) Rapidly change their patterns of concrete organisation in order to be able to continuously explore new technological options in the best way possible.

21. Even in bureaucracies in phase two, the need for occasional organisational change has generally been recognised. According to phase two patterns, the fulfilment of this need has rested with top management with or without the support of an "Organisation development department" or something similar. When we refer to new forms of organisation, however, we mean a different approach - an approach which implies the ability to re-create oneself in organisational terms is shared by the whole enterprise. It is for this reason that the issue of organisation re-creation turns, to a large extent, on the issue of what

patterns of work organisation are applied. Below, we will take a brief look at the emergence of different patterns of work organisation.

C. The emergence of new concepts of work organisation

22. For most of their existence, Taylorism and related schools of thought have been criticised. The "human relations" school, emerging out of the Hawthorne project (Mayo, 1945; Roethlisberger and Dickson, 1952) was such a criticism, in theoretical as well as in practical terms. The concepts that have dominated the last decades - job rotation, job enlargement, job enrichment and autonomous groups - are, however, somewhat newer, in that they were introduced mainly in the 1950s. These concepts are, today, very well worked through and there is a vast literature. Here, only a few points will be discussed.

23. One often encounters the view that these concepts refer to generally specified patterns of organisation with given consequences, i.e. job enrichment always leads to this; autonomous groups to that, etc. However, this is not the case. The content and implications of any of these patterns are to a large extent dependent on the purposes for which they are introduced and in what context. Job rotation, for instance, can benefit the workers if it is part of a scheme for self-administered learning and development, but can equally well be a major burden if it is introduced and controlled by management alone in order to make people more "flexible". The concept of the autonomous group is less open to such alternative meanings as it is based on a larger number of criteria (Herbst, 1962; Gulowsen, 1971). On the other hand, there is a fairly large number of different patterns of organisation that can be subsumed under this concept. This does, in fact, follow from the point made above about organisation and regeneration. If we presume that the autonomous work group is a concept positively related to learning, and that in this case learning means the continuing reconstitution of the work group, then a group can take on a number of different forms without losing its autonomy. Talking about "new forms of work organisation" in connection with such concepts as job enrichment and autonomous groups is, today, somewhat misleading, as these concepts are now about 30 years old. They are not outdated, however. They are related to issues of continuing relevance, but they must be placed in a broader context as part of a network of concepts and strategies that can lead to changes in working life.

24. When the early efforts to give such concepts practical expression were made, the focus was on industry at the shop floor level, with the idea being to change job design criteria. These efforts, such as in the British coal mines (Trist and Bamforth, 1951; Herbst, 1962), in the plants included in the Norwegian Industrial Democracy programme (Emery and Thorsrud, 1976), or the URAF programme in Sweden (Sandberg, 1982), largely took on a very local character. Changes took place among smaller groups of workers - sometimes - in smaller plants. However, the change process soon was halted in two respects. Firstly, it was difficult to move beyond the shop floor to get the enterprises to develop patterns of supervision and planning that could support the development of autonomy on the shop floor. Secondly, it proved difficult to move beyond the experimental sites and enterprises to achieve a broader process of diffusion (see, for instance, Bolweg, 1976).

D. Further expansion of "phase two" thinking

25. While the 50s, 60s and early 70s saw the emergence of alternative ideas of work organisation, diffusion was very limited. The main characteristic of the development of working life far into the 70s was the continued expansion of "phase two" thinking. Ideas about work simplification and tighter planning and control made their way into the non-industrial sectors. Visions were created of the office becoming some sort of technologically controlled system for information processing, the office clerks performing simple marginal functions. In this section, we will touch upon some of the aspects of automation in this period.

26. One of the principles of phase two is the idea that human beings have certain universal capabilities that can be ordered along one dimension. Automation can be defined as the taking over by machines of successively higher orders of human capabilities. It has, however, proved technologically difficult to really reproduce the human ability to make judgments, to actively communicate with each other, to take initiatives, etc. Hence, the automation process has not merely substituted technology for people, it has also implied a transformation of production systems to achieve the necessary prerequisites for automation. This has meant a transformation from a demand for "higher" to a demand for "lower" types of human capabilities - to automate, men and machines must, so to speak, meet halfway. This is explained in the example below.

27. When a new type of process is developed there are often many control problems. The process may show many dramatic variations and be very demanding on the operators. In terms of a flow chart it can look like this:

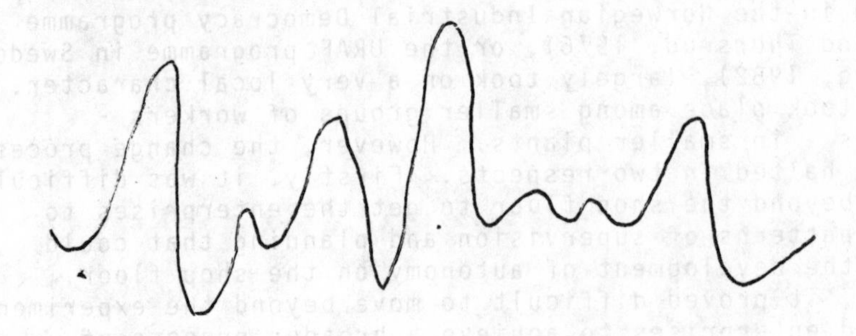

28. While this can be challenging for the operators, it can also be stressful. Consideration of the work environment as well as of productivity can show the need for reduced variation, which can be pictured something like this:

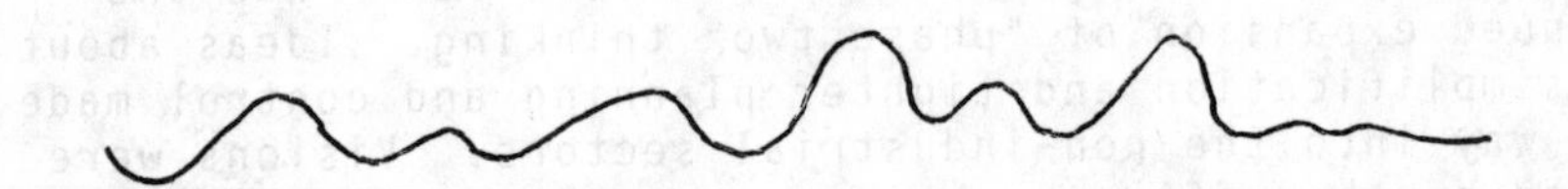

29. Here, the variations are more moderate. They provide reasonable challenges and an inspiring work environment without having the constant crises of the first stage. However, rather than stopping here, where working conditions are good from the workers' point of view, one often proceeds to a flow situation looking something like this:

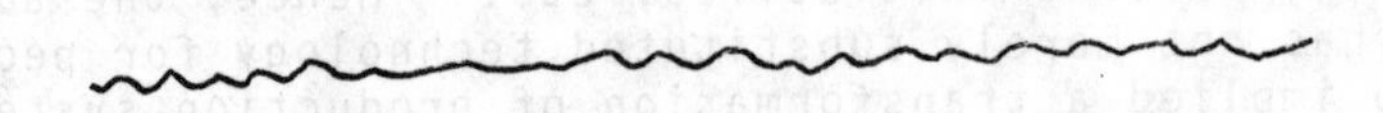

30. Here, the variations are very small and easily controllable. When this level is reached, generally after a number of years' accumulation of experience and heavy

investments in technological improvements, the stage is reached where full automation emerges as a practical possibility - or the other way around: the process is transformed to this stage as a principal step towards full automation. In this stage, however, work often takes on negative characteristics, such as monotony and boredom (Agervold and Johansson, 1983). Large parts of industrial processing is today of this type. The same is true in power plants. If full automation is achieved, the human problems disappear, but this last step has turned out to be more difficult. Most processing plants are still manned but the operators' jobs are very unstimulating, which reduces their alterness and contributes to the risk of accidents.

31. Another example of the move to phase two thinking can be found in the printing industry in newspaper production. Conflict over the use of new technology has caused confrontations between printers and managers in many countries. Studies indicate that the way in which new technology is used can cause problems. In a study of a Norwegian newspaper (Ødegaard, 1981) it was found that a working relationship characterised by interactive dependencies between the printers had been changed to one of sequential dependencies, a major transformation of the workplace. People who have work-determined interactive dependencies are necessarily involved in dialogue - work is a source of that particular human activity which is most closely related to learning and development. Sequential dependencies imply "passing messages along the line". The study also demonstrated that this change in work relationship structure was not a necessary consequence of technology. It was possible to maintain new technology and at the same time to re-constitute interactive dependencies. The change here, from interactive to sequential dependencies, exemplifies the changes in a number of other sectors of working life. Office automation is, for instance, generally developed on the basis of a vision of "information" as something which can be objectified and machine bound. The people in the office are no longer seen as the main generators and carriers of the information. Instead, they become providers of routine inputs based on clear specifications or process supervisors. This, in turn, implies a change from a situation of dialogue to a situation where the task is to see to it that "the messages flow smoothly".

32. A third example of the problems which follow in the wake of automation in combination with phase two thinking can be found in large mail-sorting stations, of which a great number have been built all over the world during the last decade. Here, great faith has been put in

technological transport systems, optical reading equipment signal systems to control the flow through of the letters, etc. However, a number of problems have emerged in these workplaces (see, for instance, Gustavsen and Hunnius, 1981; Smyth, 1982). The focus on advanced technology has resulted in a neglect of the more manual operations of which there are still quite a number. At the same time, the technologically advanced parts of the system have often malfunctioned or had practical capacity limits far below their theoretical ones, with a resulting increase in reliance on the manual operations to function as a buffer, to take peak loads, etc. The possibilities for performing these functions have, however, been impaired by neglect and by priority being given to "high technology" solutions. An imbalance has emerged in the relationship between the different components of the whole system. It is quite common for changes to be introduced through the introduction of new technology at the expense of the development and and improvement of traditional work methods. It must be recognised that automation - or rather the design of a production system - must form part of a holistic perspective, i.e. a system which functions as a totality. Some sort of balance between the different components must be maintained in order that, for example, human mail sorters do not alternate with ultra-rapid machines.

E. Renewal of the movement towards new forms of work organisation - the problem of first line supervisors

33. After a period of setback, the development and diffusion of organisation again has started to gain momentum. Developments and solutions have emerged to give some pointers and ideas concerning the organisation of the enterprise as a whole.

34. The early projects demonstrated that while there was a need for restructuring on the production floor, this could not be achieved unless other parts of the enterprise organisation were also changed. This led to the problem of what to do with the first line supervisors and other managers "close" to the production level. A high degree of division of labour demands more decision-makers at higher levels. When autonomy is developed on the shop floor, these positions become redundant. Four solutions for what to do with these positions were considered in the early projects:

(i) Eliminate the supervisors' jobs.

(ii) Change the role of the supervisors away from inward-directed management towards "boundary control" (management tasks pertaining to co-ordination with other departments, external environment, etc.).

(iii) Move the supervisors upwards, towards higher levels of management.

(iv) Move the supervisors downwards to operator level, for instance as a senior operator.

35. All these solutions presented difficulties, particularly in the short run and largely explain why experiements in new forms of work organisation were stopped in many countries. As time has passed, however, some solutions have been found which can be widely applied. For example, when a supervisor retires or is replaced, his or her role can be greatly changed in ways which would not be possible just through retraining of the existing personnel. However, more recent projects indicate that the main issue lies elsewhere. The extent to which a supervisory role is inwardly or outwardly directed is less important than the extent to which all instructions or procedures to be applied at the local level are given from the outside (Ødegaard, 1983). The question is the extent to which it is possible, within each work area, to achieve a reasonable degree of development in terms of planning and testing new solutions to problems of production and organisation. It is only if such local development is possible that the workers can participate in the generation of their own environment. Self-management of an externally determined environment is of limited value - not to say that it may be a contradiction in terms.

36. If we turn back to how the main considerations behind design of work organisation should be defined, we see that the emphasis on local development is in harmony with certain of these considerations.

- Industrial democracy can be seen as specific patterns of organisation. If this is the case, there is little need for local development. Democracy will exist if one can work in an autonomous group while democracy will not exist if this possibility is denied. If, however, democracy is defined not only in terms of specific patterns of organisation, but also in terms of who decides what patterns of organisation should prevail in the workplace, the possibilities of participation in the constitution of the workplace become of critical importance. The patterns of organisation associated with democracy will, in this case, appear as rules ensuring each and every worker a say in the process, but not as fully specified patterns of organisation. The author believes that the latter view of industrial democracy is the more correct (Gustavsen. 1983 b).

- It appears that views concerning the constituion of a good work environment not only depend on its specifications but also on who has designed them. Worker reactions to identical patterns can vary according to the degree of influence they have exerted over the definition of the patterns. This is not irrational behaviour, but expression of a deep-seated human tendency to want to be in control over one's own situation.

- A technological explosion implies that productivity must be defined as the ability of an enterprise to continuously diversify and renovate itself in organisational terms. If this process is to be efficient it must encompass the enterprise as a whole and this demands that all elements can participate in it.

F. The organisation as a whole

37. For local development and worker influence over their immediate environment to be possible, changes cannot stop at this level. Supervisory roles are imbedded in more general organisational structures determined by such factors as the type of planning systems used, the degree of centralisation or decentralisation, and the way experts are used, etc.

38. It has already been mentioned that the idea of a "one best organisation", applied throughout a large enterprise of some size for a long period, must be abandoned. Instead, a number of different patterns should be mastered and new patterns continuously created. We cannot raise and discuss all issues pertaining to alternative patterns of internal enterprise organisation here, but a few examples can be mentioned.

39. Among the various patterns of organisation that need to be developed and mastered, those patterns pertaining to development - project organisation, matrix organisation, etc. - must be given priority. In phase two, formal hierarchies were given the most attention. There are probably a number of reasons for this, one of them being that the formal hierarchy expresses the power relationships. Power does not, however, create new

products and there is now a clear need to emphasise those patterns that have bearing on such issues.

40. Secondly, patterns of developmental organisation must encompass the enterprise as a whole. That an R and D department should be organised differently from a bookkeeping department has long been recognised but this, however, is no longer the point. The development of new products can no longer take place exclusively in an R and D department. Today's technological complexity and demand for product quality mean that it is more and more difficult to make products that are not designed without guidance from the production workers. Consequently, one of the major issues confronting the more advanced enterprises in, for instance, the mechanical industry, is the utilisation of CAD/CAM systems so that all functions and all people - from production to sales - can participate actively in the development of new product families.

41. The emphasis on development-oriented patterns of organisation does not imply a neglect of all formal and stable aspects of enterprise organisation. In order to engage in developmental tasks, people generally need a certain degree of security. This must be expressed in organisational terms, for instance, in internal rules pertaining to how personnel issues are dealt with, specific legal obligations distributed among the employees, etc. A major problem about phase two patterns of organisation is that while they are not particularly well-suited to developmental tasks, they are generally not very satisfying from the perspective of internal rules and procedures either. It is far better to define what is needed for stability, security, and also for the minimum amount of formal power required and then to develop a formal structure accordingly.

42. Lastly, the relationship between "expert" competence and the type of competence acquired by doing a task, which can be called participant's competence, should be mentioned. The tradition over the last few decades has been to put great faith in experts of all types: planning, technology, economy, personnel management, etc. Clearly, highly educated professionals do have contributions to make but there are, however, strong indications that expert knowledge has been over-rated. In working life, this has emerged in several areas. Such projects as the Industrial Democracy programme in Norway brought various instances to light where the workers could solve a problem equally well or better than the engineers (Emery and Thorsrud, 1976). The competence gained through working "inside" a system is very often valuable and also unique. The same point has emerged

in Scandinavian projects pertaining to systems for planning and control such as the iron and steel project in Norway (Nygaard and Bergo, 1974) and the DEMOS project in Sweden (Sandberg, 1983). The work environment improvement efforts of the latter 70s demonstrated the same point (Gustavsen and Hunnius, 1981). However, even though some experience has been gained in the development of new patterns of organisation and relationships between experts and participants, a lot remains to be done in this field.

G. Towards a new enterprise

43. While the type of developments indicated above are at least common enough to have attracted the interest and analyses of theoreticians and practitioners for almost a couple of decades, there are also signs of a newer and still rather rare movement beyond this. For a long time there has been a trend in the ratio of supervisory to non-supervisory personnel in the direction of more and more of the first. Organisational developments, as indicated in the preceding sections, can take place without any major break with this trend even though reduced reliance on expert knowledge, together with today's cost pressure, will eventually imply a force towards fewer supervisors and staff bodies between top management and the production floor. However, a movement towards a total enterprise organisation which differs in shape from traditional ones can to some extent be identified. Patterns that can be seen as emerging - albeit in only a few cases - are as follows:

- Most important functions are located on the production floor and integrated with production.

- At the same time, the control and supervisory tasks of the managerial levels are drastically reduced.

- The growth in the number of levels between the top management and the shop floor is halted, and to some extent even reversed.

- The people remaining on the in-between levels have their tasks transformed from control to support.

- Top management grows and becomes more collective.

- Local worker representation is strengthened.

- Important characteristics of enterprise policy are formed in a direct dialogue between top management and worker representatives.

44. The cases that exhibit this pattern are often relatively special. Norwegian examples include, for instance, Skotfos, an old paper factory that went through a crisis and remained in operation because of a special agreement between management and the workers (Engelstad, 1983), and Norsk Data, a highly successful but small data-processing firm. The latter belongs to the new sectors which have a short history and few traditions. A further example is the Scandinavian Airlines System, which now pursues a conscious policy of strengthening both the parts of the organisation that directly deal with the public and top management as well. The resources to do this come from thinning out the in-between levels. It can be added that SAS is at the moment among the most successful, economically speaking, of the international airlines.

45. In terms of simple illustration, the movement from the hierarchy towards the pattern indicated here, can be pictured as a movement from the pyramid to the mushroom (even though it is a somewhat atypical mushroom, as it is broadest at the base).

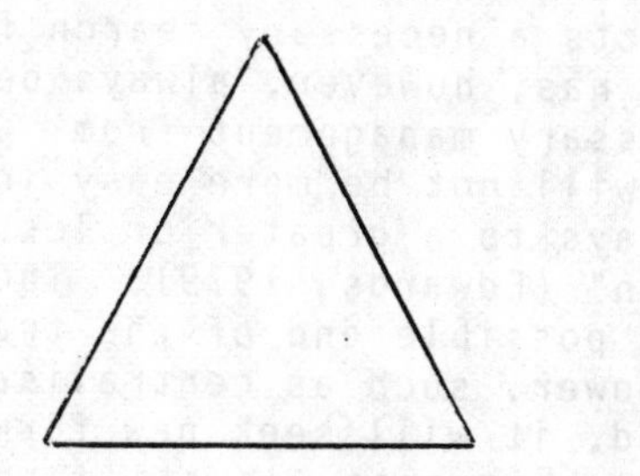

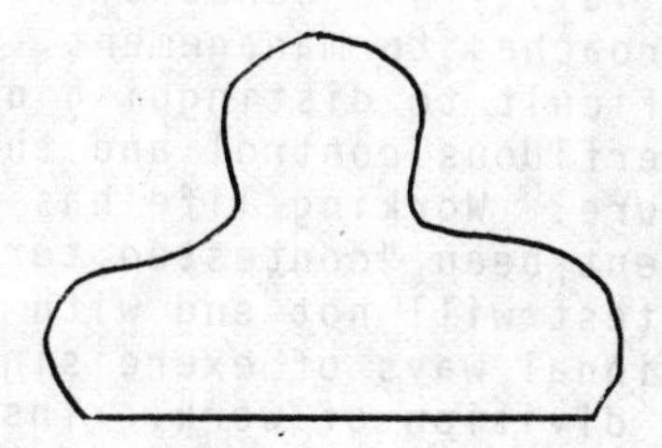

46. To the extent that this movement towards a mushroom is happening, one may wonder why it takes place. There are, of course, a number of enterprises in existence which have traditionally had very thin managerial ranks such as in the textile industry. Here, however, the shop floor has been organised in a maximally Tayloristic way. The mushroom shape, as a response to new technologies, for example, is due to different reasons. One point seems to be the following: if an enterprise is to develop a number of different organisational patterns and to continuously change them, there is a need for very good and undisturbed communication between workers and management because the

diversification and changes raise a great number of issues which will need to be clarified and settled. In other words, at a higher level of abstraction the meta-level necessary to develop and sustain this organisational pluralism includes communication (Gustavsen, 1983 b). This is in line with the position of the German philosopher Habermas who underlines communication, or dialogue, as the chief generative process among human beings. Phase two organisational patterns do not work very well, however, from this perspective. Communication is often disturbed, skewed, irrational and time-consuming. Immense amounts of power must often be wielded for a phase two hierarchy to really react rapidly to changes. Then, however, there will be no possibility for collective developmental efforts.

47. The "mushroom" type of enterprise pattern implies that the traditional control mechanisms - a well-developed hierarchy of middle and lower managers and a high degree of division of work on the shop floor (Braverman, 1974) - recede into the background. From a control perspective, a vacuum is created. At the moment, there are theories which suggest how new control mechanisms can be developed to replace traditional ones. Concepts such as symbolic management, enterprise culture, etc., attract interest. Selznick's "Leadership in Administration" (Selznick, 1957) which deals with the conscious creation of organisational values to direct and control behaviour, experiences a revival. This tendency reflects a necessary search for new approaches to management. It has, however, always been difficult to distinguish necessary management from superfluous control and this will not be more easy in the future. Working life has always to a greater or lesser extent been "contested terrain" (Edwards, 1979). The contest will not end with the possible end of the traditional ways of exercising power, such as centralisation and division of work. Instead, it will seek new forms. What is going to happen within this area, is still an open question. What is clear, is that the now emerging forms of enterprise organisation will pose new problems over a very broad range of issues and that these problems will come to dominate thinking and development over the coming decades.

H. The issue of moving directly from pre-tayloristic to post-tayloristic patterns of organisation

48. In a number of countries, particularly in those that are not fully industrialised, there is a growing interest in the possibilities of moving from a pre-tayloristic stage without passing through fifty years of

hierarchisation and division of work. When the concept of the autonomous group was originally introduced as an advanced concept of organisation, its close relationship to pre-tayloristic forms of organisation, such as the traditional work groups in the building industry, in forestry, etc., was noted (e.g. Gulowsen, 1971). This concept allows a transition directly from a pre-tayloristic to a post-tayloristic phase or - in terms of the concepts used above - from phase one to phase three. In so far as enterprise characteristics like those indicated above emerge out of a need to deal with advanced and rapidly changing technology, the same perspective may be applied. The "mushroom" is similar to what is found in many relatively less industrialised areas: a few middle managers; direct dialogue between top management and the workers; many functions allocated to the shop floor, such as maintenance or sometimes the hiring and firing of workers or the development of production methods, etc. Hence, there may be another bridge between pre-tayloristic and post-tayloristic patterns of organisation beyond the one constituted by autonomous groups.

49. The possibilities of bypassing phase two are more dependent upon political conditions than upon conditions controlled by the enterprise. A post-tayloristic pattern of oganisation has to be based on active, competent workers with a reasonable degree of self-management. It is obvious that people who are expected to be active and competent in certain respects, cannot be passive and submissive in others. Hence, the overall patterns of organisation in society must be such that they accept and even encourage local responsibility and active political influence from the grass-roots. Furthermore, many of the skills and competence needed in working life have to be collective: the work must be carried out by groups or larger systems of people. When collective action is encouraged in daily work, one cannot expect people to refrain from collective action in other relationships such as, for instance, those expressed through unions.

III. Means of improvement

50. The question of means of improvement is identical to that of strategy for reform which is very complex. Below some of the dimensions relevant to this issue are discussed.

51. Firstly, there is generally a need to apply more than one single parameter. There are numerous cases where researchers, consultants, and others have claimed to have found the key to a transformation of working life; currently "quality circles" is the popular concept of many "single parameter reformists". It is very rare, however, for a single means to achieve very much. An exception is when the situation has become ripe because a number of conditions have changed in appropriate ways before the single parameter is applied. Then it can have an impact and sometimes foster the belief that it was the single parameter that did the trick. However, there is practically always a need to apply a number of means, which is why the concept of strategy must be used.

52. Secondly, as pointed out above, there has been a development within the work reform movement from relatively local changes towards efforts at dealing with the enterprise organisation as a whole. The introduction of autonomous groups in a few experimental sites is a somewhat different task from that of trying to get a relatively broad number of enterprises to generale new patterns of organisation throughout all their operations.

53. Thirdly, while there has been a shift in emphasis from changes on group and workshop level to changes on the level of the organisation as a whole, there has also been a development concerning the overall diffusion process. Here, one can talk about stages and, using Norway as an example, the following can be identified (Emery and Thorsrud, 1976; Bolweg, 1976; Engelstad and Ødegaard, 1979; Gustavsen and Hunnius, 1981).

Stage 1 (1965-1970): Introduction of ideas about alternative forms of organisation. Demonstration projects. Some efforts at diffusion.

Stage 2 (1970-1975): Problems and setbacks. Confusion rather than diffusion. Defensive measures to avoid reversals within the first experimental sites.

Stage 3 (1975- ..): The diffusion process starts to pick up. A slow but steady increase in number of relevant developments.

Stage 4 (1975-1980): Political efforts, such as legislation, the use of public resources etc., to influence patterns of organisation in working life.

Stage 5 (1980 - ..): The main organisations in working life - the Federation of Trade Unions and the Employers' Confederation - take on a broader responsibility for work organisation development. New agreements, new patterns of work and organisation within these institutions.

54. The times given in this list of sequential events are approximations. They pertain to Norway, which is a country where this type of development started relatively early. The pattern indicated here is, however, not peculiar to Norway. The setback after the demonstration period seems, for instance, to be rather general. Political efforts have emerged at one stage or another in a few countries, even though legislation pertaining directly to organisation of work is rare. The clearest examples of such legislation are the Norwegian Work Environment Act Article 12 (Gustavsen, 1977; Gustavsen and Hunnius, 1981) and the Working Environment Act of 1980 of the Netherlands (see, for instance, ILO, 1983). Sweden follows roughly the same pattern as Norway. The Federal Republic of Germany illustrates many of the same elements but in a somewhat different order. For instance, its major public effort, the "Joint action programme on research to humanise work", was launched before any demonstration phase had been held. Hence, a number of the issues which are usually dealt with in the first phase, and some of the ensuing difficulties which emerged within the framework of the programme itself, are probably reasons for the many debates about this programme (Freiderich Ebert Stiftung et al.,1982). The chief point, however, is that the means most appropriate to the furthering of a development towards new patterns of organisation will to some extent depend upon the stage one is at. The task of launching "high visibility" demonstration projects to show society that there are alternative ways of organising people at work, is a separate task from overcoming the setback that almost invariably seems to follow "high visibility" demonstrations. Changes in working life are most efficiently achieved if:

- A number of means are used.

- The means are linked together in an overall strategy.

- This strategy is built on the recognition that the different levels - the workshop, the enterprise, the main organisation in working life, legislation, etc. - must support each other. Change depends, to a great extent, on the degree of harmony one can ensure between means of different types and at different levels.

- This strategy is also built on the recognition that changes in working life pass through a sequence of different stages. Therefore the peculiarities of each and every stage must be reflected in the strategy.

IV. Themes and priorities concerning future action

55. We have underlined that any change in working life necessitates a multi-level and multi-means approach with the various elements linked together in a strategy. Mutually supportive development must take place within individual work roles; within the workgroup; the factory; the enterprise, and possibly within the community, employer and employee organisations at various levels; as well as in social institutions such as legislation. Among the many issues and problems involved, some are better covered than others by experience and research.

56. Among the better covered aspects are changes at the workplace and enterprise level. Experience in this area has accumulated over several decades. Some of the main topics include, firstly, the problem of where a change should start - on the shop floor, at top management level, or somewhere in-between. This problem emerged in early experiences such as the Glacier project (Jaques, 1951) and has been broadly discussed in relation to a number of other projects (e.g. Hill, 1976). The socio-technical school has generally given preference to a "bottom-up" approach, while the organisation development school has generally preferred to start with top management. Today's view on this issue is generally that the starting place should depend on local conditions. Efforts emerging out of considerations of democracy and work environment will always lean towards a shop floor approach.

57. A second dimension where differing views can be found concerns what constitutes the chief engine of change processes. In some instances there is a strong belief in technological development. This seems to be the case in the Humanisation programme in the Federal Republic of Germany where efforts at creating new patterns of

organisation are strongly linked to technological development. Others have emphasised changes in people, such as a higher level of education, which is thought to create a demand for more rewarding and stimulating work. A number of other factors can be brought forth as possible leading elements in change processes, or as major conditions affecting the possibilities for change. This author does not subscribe to any "single parameter" theory in this respect. The conditions and their engine will vary from country to country, over time, and even from enterprise to enterprise within the same country. General factors, such as technological development, are important but, as pointed out above, not in isolution from a number of other conditions such as those which shape the organisational ideas controlling the design and use of technology.

58. A third topic concerns the change agent: is it necessary to have one or a few persons to bear a leading role in the change process and who should they be? Here, we confront the issue of the use or non-use of external resource persons such as consultants or researchers and the type of project organisation to be developed, etc. Below, we make comments on research, with particular emphasis on social research.

59. Social research - or rather, interdisciplinary research with social aspects as a main element - has clearly been important in the development of new patterns of organisation in working life. Social research has been involved in most of the significant developments within this field. The importance of social research has, furthermore, increased rather than decreased over the last decades as more and more emphasis has come to be placed on alternatives to the "social engineering" implicit in Taylorism and traditional theories of administration. This notwithstanding, the role of social research today is under heavy debate in a number of countries and is, if anything, on the decline. For instance, the current revisions of the Humanisation programme in the Federal Republic of Germany seem to be in this direction. This exemplifies a common trend though there are exceptions. In Sweden, for example, research has played a modest role in actual workplace developments since the conclusion of the early projects in the 70s but this could change in the future as research policy may more strongly emphasise the need to give support to concrete workplace developments. In Norway, research has had a slow but steady increase in its influence and position. After the Industrial Democracy Project it came to play an important part in the development of the work environment reform in the latter 70s (below) and has recently been invited to join the board established

by the Federation of Trade Unions and the Employers' Confederation to supervise and support efforts under a new agreement on enterprise development.

60. The problems currently facing social research are due to a number of different circumstances, some of them clearly of a political nature about which little can be done. However, one source of difficulty concerns the role of research and the large amount of confusion that has emerged around the concept of "action research". It is often thought that "action research" means that the researchers go into the workplace and start directing changes and if the directives are not acted upon, they publish a critical report. Since this is a very common belief and is, to some extent, borne out in research practice, it is worth pointing out that today "research in action" implies a number of different roles and options that distribute influence between researchers and those with whom they collaborate in a number of different ways:

(i) When research first started to become involved in workplace developments, it tried to take on a leading role. This was due to various circumstances, not least a belief among researchers in general theories which were supposed to enable the researchers to make autonomous interpretations of workplace problems and develop strategies for change accordingly. In the period of the "high profile" demonstration experiments, this was very much the self-definition of research (Gustavsen, 1983 b). However, this is no longer the case.

(ii) A participative model has emerged, in which influence over the definition of problems and development of solutions is shared equally between researchers and the people in the workplaces where the changes are to occur.

(iii) A supportive model has also been developed in which research acts as a resource to be called upon by those concerned when necessary but where research does not take its own initiatives.

(iv) A fourth role can also be identified in which research takes responsibility for the "infrastructure" of development: those areas or settings which are called for if new ideas are to be created, but where it is left to those directly concerned to utilise these arenas (Ødegaard, 1983).

61. All these models, and combinations of them, are available to research in actual workplace development (provided that there are researchers available who master these various models for participation in workplace development). Given the still heavily descriptive/interpretative direction of social research, this may not be the case in a number of countries.

62. While the enterprise level can be said to be at least relatively well covered in terms of experiences and research, this is not the case when we look at the other ingredients of change strategies. If it is accepted that significant changes in working life beyond "the odd exception" or "the spectacular show case" are dependent upon a supportive superstructure, these issues become of critical relevance. Here, we will look into one such issue: the use of legislation (these arguments can also be applied to centrally defined agreements and other general normative systems).

63. Public regulation can be perceived in various ways. A somewhat naive, but still important, way is to see it as a machine-like type of process which starts with the politicians drawing up clear goals defining what they want to achieve, which are then converted into unequivocal legislation. This legislation, in turn, forms the basis for the development of supplementary specifications, inspection systems and specific corrective measures. Through these various means, "reality" is successively brought to correspond to the clear and logically structured goals of the political level. "Reality", however, is often resistent. Below, a brief picture will be given of what seems to be the relationship between ideals and realities in public reform efforts. The points were made in connection with a discussion and development of strategy within the field of workplace health and safety in Norway (Gustavsen and Hunnius, 1981; Gustavsen, 1982). They are somewhat general and similar discussions have emerged in other countries, such as Italy (Assenato and Navaro, 1980) and the province of Saskatchewan in Canada (Saas, 1979).

IDEAL	REALITY
Political:	
- Clear goals	- Diffuse goals
- Logically structured	- Often inconsistent
- Accompanied by cost-benefit analyses	- Vague ideas about costs and benefits
- Conflicts (e.g. between such values as health and economy) are settled	- Conflicts are not settled
Administrative:	
- Supplementary rules are issued rapidly	- Supplementary rules come late, it at all
- Supplementary rules are "technical matters" as goals and conflicts are already settled	- Supplementary rules create conflicts and demand large resources for their settlement
- Most work environment problems are visible	- Most work environment problems are more or less hidden
- and can be brought to light through brief visits from inspectors or other "experts"	- and have to be "dug forth" through a stepwise effort over time
- Most problems are covered by existing solutions	- There is a great need for new solutions
- Most workplaces are reached by the public inspection system, such as the labour inspection	- Most workplaces are not reached by the public control system
- Inspection priorities are decided on the basis of the amount and seriousness of problems	- Inspection priorities are mostly decided by requests (workplace activity independent of the public efforts)
- Changes are ordered	- Changes are advised
- Most breaches of rules are followed by sanctions	- Few breaches of rules are followed by sanctions

Research:

- Uncovers "laws"	- Rarely uncovers "laws"
- Which can be converted into workplace specifications	- Workplace specifictions even more rare
- through objective studies	- Objectivity at best a difficult concept
- which can be performed independently of the action situations	- Research and action often interdependent
- and lead to a cumulative generation of knowledge	- Cumulation limited by frequent shifts between - and changes of "paradigms"
- that exist "autonomously" (independent of particular situations or contexts)	- Knowledge is often of a "local character"

64. Empirical support for what is termed "reality" here is now quite broad and ranges from general studies of public administration to specific studies of health and safety at work. In Norway, the large discrepancy between ideals and realities concerning policies and legislation in working life emerged in the 1950s in a study of the working conditions of household servants (Eckhoff, Aubert and Sveri, 1959).

65. Cost-benefit analyses often demanded by politicians are impossible given the present stage of development of economic concepts and categories. These do not allow us to isolate the work environment component of, for example, investments in new technology. Altogether there are five studies on labour inspection, four in Norway (Karlsen et al., 1975; Halgunset and Svarva, 1980; Gustavsen et al., 1981; Karlsen et al., 1982) and one in Sweden (Lundberg, 1982) which all demonstrate that workplace visits are scarce and priorities are largely decided according to demand. This is not a criticism of this particular branch of the public sector, just a statement of restrictions on what can be overcome. The Scandinavian countries are fairly well-covered by inspectors, measured by the number of inspectors in relation to the number of workplaces they have to cover (for some further discussion, see Gustavsen, 198 ; 1983).

66. The argument indicates that limited trust should be placed in public efforts, at least if such efforts are structured according to the conventional legal-administrative model. This does not imply an argument for the abolishment of public efforts, nor even for de-regulation in a conservative sense, but it does imply a need to seek an alternative prime vehicle in the reform process. Legislation as such and workplace inspectors, occupational health services, etc., are not able to act as the chief force in broad reforms in working life. Where, then, can such a force be found?

67. There is only one place to go: to the workplaces where the changes are to take place. To utilise such local resources, however, it is necessary to recognise that legislation and other elements in the "superstructure" must be designed so that local action becomes possible and is encouraged. One cannot expect local efforts to emerge easily under legislation which introduces heavy demands for expert evaluations or time-consuming demands for public sanctioning. When strategy is designed, it is also important not to forget that in all countries the great majority of employees are in small enterprises - often without a single "expert" on any topic whatso ever. One way of approaching the issue of legislation in support of local development is to introduce a set of broadly defined rights and duties, for instance as follows (Gustavsen and Hunnius, 1981):

- Firstly, a broad duty to act within the field of health and safety pertaining to both parties, but particularly to management.
- Secondly, a broad right to raise health and safety issues. This right also pertains to both parties but particularly to employees.
- Thirdly, the right to a reasonable degree of worker participation.
- Fourthly, liability to public sanctions if these broad duties are not taken seriously. Or, in other words, the power wielded by the public system should be used to force local activity and not to correct specifics.
- Fifthly, if the broad duty to be active within the field is fulfilled, the individual enterprise has wide latitude for defining what the problems are and what solutions should be developed and applied. There is, in other words, a version of a "freedom within obligation" principle.

68. One country where the legislation within the work environment field is now relatively close to these principles is Norway. When the present act was developed in the 1970s, the issue of reform strategy was still an open one and the act reflects different strategies. However, the later guidelines and interpretations developed by the Ministry of Labour (such as KAD, 1981) have moved more and more towards the local development line.

69. There is, of course, not only a need to introduce certain broad principles. Various, more specific rules must also exist in order to make decisions "accessible" at the local level such as rules pertaining to evaluation of evidence, burden of proof, concepts of causality, etc. (a review of some of these issues can be found in Gustavsen, 1980 c). There is no question of abandoning traditional institutions, such as the labour inspection, but to give them a new role: as initiator and supervisor of development, rather than as a body that issues many legally binding demands for change which generally pertain to very small technical details.

70. The points mentioned here do not necessarily indicate the only possible way to structure political, legislative, and public efforts so that they can interact in a positive way with local development. A different path was taken in Sweden where the 1976 Co-determination Act (Medbestämmandelagen) focuses on negotiations between the different groups. This is a very traditional approach though the Co-determination Act also ensures that very broad rights and possibilities are given to the workers and emphasises issues beyond wage negotiations. This legislation stresses that organisation of work should be subject to negotiations and agreements. The Co-determination Act, however, developed along principles not unlike those mentioned above, i.e. management has a broad duty to raise issues with the employees before changes are made; the employees have a broad right to take up issues and demand negotiations with management. The Act concerns issues relevant to worker participation (e.g. organisation of work). Public sanctioning can be applied if the rules are not applied, and - last but not least - local agreements tailored to fit local conditions are possible.

71. Neither the Co-determination Act in Sweden nor the Work Environment Act in Norway have been unqualified successes in the short run (Hammarström, 1980; Gustavsen, 1982) but this could hardly be expected. It will take much more time to break existing bureaucratic patterns of mediation between the political apparatus and the people concerned which exists in most societies to day. Reforms such as

these indicate some of the issues one has to face and at least some of the options available for how to deal with them. In the future, the creation of total structures in support of development on the level of society as a whole will be the critical issue. If by nothing else, they will be required by economic conditions, which will demand rapid increases in ability to innovate, to produce at reasonable quality levels. Clearly a precondition for this is good labour-management relationships which can only be achieved through letting developmental efforts emerge from a common ground constituted by productivity, democracy and work environment.

Note:

[1] Director, Institute for Work Psychology, Work Research Institutes, Oslo, Norway.

References

J.D. Adams (ed): Theory and method in organization development: An evolutionary process (Arlington, NTL Institute, 1972).

M. Agervold and G. Johansson: Stress og belastning blandt procesovervågere. Sammenfattende rapport fra Nordstress-procesovervagningsgruppen (Arhus, Psykologisk Institut, 1983).

S. Agurén, R. Hansson and K.G. Karlsson: The Volvo Kalmar Plant (Stockholm, The Rationalisation Council SAF-LO, 1976).

S. Agurén and J. Edgren: New factories (Stockholm, The Swedish Employers' Confederation, 1979).

G.H. Amber and O.S. Amber: Anatomy of automation (Englewood Cliffs (New Jersey), Prentice Hall, 1962).

C. Argyris: Integrating the individual and the organization (New York, Wiley, 1964 a).

C. Argyris: "T-groups for organizational effectiveness", in Harvard Business Review (Boston, Harvard University Press) 1964 b, No. 46, pp. 60-74.

G. Assennato and V. Navarro: "Workers' participation and control in Italy: the case of occupational medicine", in International Journal of Health Services (England, Confederation of Health Service Employees), 1980, Vol. 10, No. 2.

W.G. Bennis: "A new role for the behavioural sciences: effecting organizational change", in Administrative Science Quarterly (Ithaca, Cornell University Press), 1963, No. 8, pp. 125-165.

W.G. Bennis: Organization development: Its nature, origins and prospects (Massachussetts, Addisson-Wesley OD-series, 1969).

T.O. Bergo and K. Nygaard: Planlegging, styring og databehandling (Oslo, Tiden, 1974).

R.R. Blake and J.S. Mouton: "The induction of change in industrial organizations", in Scientific Methods (Austin), 1962.

R.R. Blake and J.S. Mouton: Building a dynamic corporation through grid organization development (Massachussetts, Addison-Wesley OD-series, 1969).

R. Blauner: Alienation and freedom (Chicago, University of Chicago Press, 1964).

J.F. Bolweg: Job design and industrial democracy: The case of Norway (Leiden, Martinus-Nijhoff, 1976).

H. Braverman: Labor and monopoly capital (New York, Monthly Review Press, 1974).

J.R. Bright: Automation and management (Cambridge, Harvard University Press, 1958).

T. Burns and G.M. Stalker: The management of innovation (London, Tavistock Publications, 1961).

T.H. Conant and M. Kilbridge: "An interdisciplinary analysis of job enlargement: technology, costs and behavioural implications", in Industrial and Labor Relations Review (London, Eclipse Publication Ltd.), 1965, No. 18, pp. 377-395.

L. Davis and A. Cherns (eds): The quality of working life (London, The Free Press, 1975).

Deutsche Bundestag: Drucksache 10/6, 10. Wahlperiod, 060483, Bonn (1983).

T. Eckhoff, V. Aubert and K. Sveri: En lov i søkelyset (Oslo, Akademisk Forlag, 1952).

R. Edwards: Contested terrain: The transformation of the workplace in the twentieth century (New York, Basic Books, 1979).

M. Elden: "Democratization and participative research in developing local theory", in Journal of Occupational Behaviour (England, Wiley and Sons Ltd.), 1983, Vol. 4, No. 1.

F.E. Emery (ed): Systems thinking (Harmondsworth, Penguin, 1969).

F.E. Emery and E.L. Trist: "The causal texture of organisational environments", in F.E. Emery (ed): Systems thinking (Harmondsworth, Penguin, 1969).

F.E. Emery: "Searching for new directions", in M.E. Emery (ed): New ways for new times (Canberra, Centre for Continuing Education, Australian National University, 1976).

F.E. Emery and E. Thorsrud: Democracy at work (Leiden, Martinus-Nijhoff, 1976).

F.E. Emery: Towards a new paradigm of work (Canberra, Centre for Continuing Education, Australian National University, 1978).

M. Emery (ed): New ways for new times (Canberra, Centre for Continuing Education, Australian National University, 1976).

P.H. Engelstad: Skotfos mot strømmen (Skiens Naeringsråd, 1983).

P.H. Engelstad and B. Gustavsen: "Mot en ny bedrifts-organisasjon - 15 år etter", in J.F. Blichfeldt and T.U. Qvale (eds): Teori i praksis (Oslo, Tanum-Norli, 1983).

P.H. Engelstad and L.A. Ødegaard: "Participative redesign projects in Norway: summarising the first five years of a strategy to democratise the design process in working life", in Working with the quality of working life (Leiden, Martinus-Nijhoff), 1979.

H. Fayol: General and industrial management (London, Pitman and Sons, 1949).

M. Frankenhäuser and B. Gardell: "Underload and overload in working life: outline of a multidisciplinary approach", in Journal of Human Stress (Shellburne, Opinion Publishing), Sept. 1976.

Friederich Ebert Stiftung et al.: En program und seine Wirkungen (Frankfurt, Campus, 1982).

B. Gardell: Arbetsinnehall och livskvalitet (Stockholm, Prisma.LO, 1976).

D.M. Gordon, R. Edwards and M. Reich: Segmented work, divided workers (Massachussetts, Cambridge University Press, 1982).

L.H. Gulick and L. Urwick (eds): Papers on the science of administration (New York, Institute of Public Administration, 1937).

J. Gulowsen: Selvstyrte arbeidsgrupper (Oslo, Tanum, 1971).

B. Gustavsen: "De ansattes medbestemmelse i offentlig virksomhet. St. meld. nr. 28, 1976-77", in Kommunal - og Arbeidsdepartmentet (Oslo), 1976.

B. Gustavsen: "A legislative approach to job reform in Norway", in International Labour Review (Geneva, ILO), 1977, Vol. 115, No. 3.

B. Gustavsen and E. Ebeltoft: "Muljøproblemer, utviklingsarbeid og prosjektorganisasjon", in B. Gustavsen, S. Seierstand and A. Ebeltoft: Hvordan skal vi gjennomføre Arbeidsmiljøloven? (Oslo, Tiden, AOF, 1978).

B. Gustavsen and Ø. Ryste: "Democratization efforts and organizational structure: a case study", in A.R. Negandi and B. Wilpert (eds): Work organization research: American and European perspectives (Ohio, Kent State University Press, 1978).

B. Gustavsen: "From satisfaction to collective action: trends in the development of research and reform in working life", in Economic and Industrial Democracy (London, Sage Publishings Ltd.), 1980 a, Vol. 1, No. 2.

B. Gustavsen: "Legal-administrative reforms and the role of social research", in Acta Sociologica (Oslo, Scandinavian Sociological Association), 1980 b, Vol. 23, No. 1.

B. Gustavsen: "Improvement of the work environment: a choice of strategy", in International Labour Review (Geneva, ILO), 1980 c, Vol. 119, No. 1.

B. Gustavsen and G. Hunnius: New patterns of work reform: the case of Norway (Oslo/New York, Oslo University Press/ Colombia University Press), 1981.

B. Gustavsen, J.I. Karlsen and T. Pape: Gjennomføringen av arbeidsmiljøreformen i jern - og metallbedrifter i Oslo (Oslo, Arbeidsforskningsinstituttene, 1981).

B. Gustavsen: "Regulaing organisation of work: the Norwegian example", in IILS: Changing perceptions of work in industrialised countries: Their effect on and implications for industrial relations (Geneva, ILO), 1983, Research Series No. 77, pp. 85-104.

B. Gustavsen: "The Norwegian work environment reform: the transition from general principles to workplace action", in C. Crouch and F. Heller (eds): Organizational Democracy and Political Processes (Chichester, Wiley, 1983 a).

B. Gustavsen: Sociology as action: On the constitution of alternative realities (Oslo, Work Research Institutes, 1983 b, draft).

J. Habermas: Theorie des Kommunkativen Handelns (Frankfurt a.M., Suhrkamp, 1981).

J. Halgunset and K. Svarva: Arbeidstilsynet - politi eller rådgiver (Trondheim, Institutt for Industriell Miljøforskning, 1980).

O. Hammarstrøm: "Medbestställmande 1977-1979 - en øversikt/ Medbestämmandeførhandlingarne på SAF-LO-PTK-omradet", in Arbetslivscentrum: Tre år med MBL (Stockholm, Liber), 1980.

P.G. Herbst: Autonomous group functioning and exploration in behaviour theory and measurement (London, Tavistock Publications, 1980).

P.G. Herbst: Alternatives to hierarchies (Leiden, Martinus-Nijhoff, 1976).

P. Hill: Towards a new philosophy of management (London, Gower Press, 1971).

ILO: Work organisation and the introduction of new technology: a survey of legislation and collective agreements in industrialised countries, Working paper for the Meeting of Experts on Automation, Work Organisation, Work Intensity and Occupational Stress, Geneva, 1983 (Geneva, doc. MEAW/1983/D.1; mimeographed).

E. Jaques: The changing culture of a factory (London, Tavistock Publications, 1951).

D. Jenkins: The age of job design (European Association of Personnel Management Institutes, 1983).

KAD: Oppfølgningen av Arbeidsmiljølven (Oslo, Kommunal-og Arbeidsdepartmentet, 1981).

R. Karasek: "Job socialisation and job strain: the implications of two related psychosocial mechanisms for job design", in B. Gardell and G. Johansson (eds): Working life: a social science contribution to work reform (London, Wiley, 1981).

J.E. Karlsen: Arbeidsmiljø og vernearbeid (Oslo, Tanum, 1975).

J.I. Karlsen, T. Pape and B. Gustavsen: Gjennomføringen av arbeidsmiljøreformen i et utvalg industribedrifter innen LO-NAF-området (Oslo, Arbeidsforsknings instituttene, 1982).

A. Kornhauser: Mental health of the industrial worker (New York, Wiley, 1965).

P.R. Lawrence and J.W. Lorsch: Organization and environment (Cambridge, Harvard University Press, 1967).

H.J. Leavitt: "Applied organizational change in industry: structural, technological and humanistic approaches", in J. March (ed): Handbook of organizations (Chicago, Rand McNally, 1965).

R. Lindholm: Towards a new world of work (Stockholm, Swedish Employers' Confederation, 1979).

L. Lundberg: Från lag till arbestmiljø (Lund, Liber, 1982).

A.H. Maslow: "A theory of human motivation", in Psychological Review (Washington, American Psychological Association), 1943, Vol. 50, pp. 370-396.

A.H. Maslow: Motivation and personality (New York, Harper, 1954).

E. Mayo: The social problems of an industrial civilization (Cambridge, Harvard University Press, 1945).

H. Mintzberg: The structuring of organizations: a synthesis of the research (Englewood Clifss (New Jersey), Prentice Hall, 1979).

H. Mintzberg: Structuring in fives. Designing effective organizations (Englewood Cliffs (New Jersey), Prentice Hall, 1983).

N.T. Nilsson: A network approach to assembly automation (Gothenburg, Chalamers University of Technology, Department of Textile Technology, 1983).

L.A. Ødegaard: Tilbake til det typografiske fag (Oslo, Arbeidsforksningsinstituttene, 1981).

L.A. Ødegaard: Deltakende handingsforskning: Lokale perspektiver på samfunnsforskningen (Oslo, Arbeidsforskningsinstituttene, 1983).

C. Pateman: Participation and democratic theory (Cambridge, Cambridge University Press, 1970).

T.J. Peters and R.H. Waterman, Jr.: In search of excellence (New York, Harper, 1982).

Projektgruppe Automation und Qualifikation: Theorien über Automationsarbeit (Hamburg, Argument-Verlag, 1978).

D.S. Pugh, D.J. Hickson, D.R. Hinings and C. Turner: "Dimensions of organization structure", in Administrative Science Quarterly (Ithaca, Cornell University Graduate School of Business and Public Administration) 1968, Vol. 13, pp. 65-105.

D.S. Pugh, D.J. Hickson, D.R. Hinings and C. Turner: "The context of organization structures", in Administrative Science Quarterly (Ithaca, Cornell University Graduate School of Business and Public Administration), 1969, Vol. 14.

F.S. Roethlisberger and W.J. Dickson: Management and the worker (Cambridge, Harvard University Press, 1952).

L. Rohlin: Organisationsutveckling (Lung, Gleerup, 1974).

T. Sandberg: Work organization and autonomous groups (Lund, Liber, 1982).

A. Sandberg (ed): Forskning før førändring (Stockholm, Arbetslivscentrum,1981).

A. Sandberg: "Trade union-oriented research for democratization of planning in working life - problems and potentials", in Journal of Occupational Behaviour (England, Wiley and Sons), 1983, Vol. 4, pp. 59-71.

R. Sass: "The underdevelopment of occupational health and safety in Canada", in W. Leiss (ed): Ecology versus politics in Canada (Toronto, University of Toronto Press, 1979).

S.E. Seashore and T.D. Taber: "Job satisfaction indicators and their correlates", in A.D. Biderman and T.F. Drury (eds): Measuring work quality for social reporting (New York, Halstead, 1976).

H. Selye: Stress in health and disease (London, Butterworths, 1976).

P. Selznick: Leadership in administration: a sociological interpretation (New York, Harper, 1957).

D.S. Smyth: "The relationship between size and performance of mail sorting offices", in Human Relations (United Kingdom, Tavistock Institute of Human Relations), 1982, Vol. 35, No. 7.

F.W. Taylor: The principles of scientific management (New York, Harper, 1913).

E. Thorsrud: "Democratization of work as a process of change towards non-bureaucratic types of organization", in G. Hofstede and M. Kassem (eds): European contributions to organisation theory (Assen, Van Gorcum, 1976).

E. L. Trist and K.W. Bamforth: "Some social and psychological consequences of the longwall method of coal-getting", in Human Relations (United Kingdom, Tavistock Institute of Human Relations), 1951, No. 4.

E.L. Trist, G.W. Higgin, H. Murray and A.B. Pollock: Organizational choice: capabilities at the coal face under changing technologies - the loss, rediscovery and transformation of a work tradition (London, Tavistock Publications, 1963).

C.R. Walker and R.H. Guest: The man on the assembly line (Cambridge, Harvard University Press, 1952).

D.M. Wells: "Unionists and 'Quality of Working Life Programmes'" (Rexdale, Ontario, Humber College, Center for Labour Studies), 1983.

J. Woodward: Industrial organisation: Theory and practice (Oxford, Oxford University Press, 1965).

AUTOMATION AND WORK ORGANISATION: POLICIES AND PRACTICES IN COUNTRIES WITH CENTRALLY PLANNED ECONOMIES

L. Héthy[1]

I. The social-economic context of automation and work organisation

1. Much research has suggested that automation is often accompanied by acute social and psychological problems such as increased insecurity of employment, "enforced obsolescence of skills" (Merton, 1967), "the loss of public identity of the job" (Roethlisberger, referred to by Merton, 1967), repetitive tasks and boredom and monotony originating from them, especially on assembly lines (Walker and Guest, 1952; Friedmann, 1956), as well as "shrinking autonomy" of the worker and his "social isolation". On the other hand, there are indications that automation offers opportunities for elimination of monotonous, dirty and physically hard jobs; for reducing differences between manual and intellectual work; and for offering more highly skilled jobs to the emerging educated labour force (Ember, tudomány, technika, 1977 [Man, science, technology]). The purpose of this paper is to describe and analyse these phenomena as they relate to the centrally planned economies.[2]

2. In considering the social consequences of automation and work organisation (including their impact on job content and work intensity), certain economic and social characteristics play an important role, namely:

(a) In all centrally planned economies, utmost importance has been placed on the specific policies of the central agencies to make the best use of the advantages of technological progress and to prevent, neutralise or counterbalance its negative concomitants that often make workers the victims rather than the benefactors of technological progress. Among such measures, which will be discussed in more detail in a later section of the present analysis, policies to maintain full employment deserve special attention.

(b) Industrialisation in most of these countries (except for the German Democratic Republic and certain parts of Czechoslovakia) has occurred fairly recently and has been based, to a large extent, on the exploitation of labour reserves. In Hungary, for example, during the period between 1949 and 1975, the share of industrial employment increased to about 40 per cent, while the number of workers almost doubled, from about one million to two million people. This growth stopped in the mid-seventies (Héthy-Makó, 1981). In the process, the introduction of labour saving techniques, including the most well-known types, such as assembly lines and highly automated equipment, has been fairly restricted. In Hungarian industry in 1979, for example, the proportion of manual jobs amounted to about 50 per cent, while that of operator jobs to only 23 per cent. Automated jobs were below five per cent, while jobs in "line" organisation, including assembly lines, amounted to 10 per cent. Automation was even more limited in other sectors, such as commerce, banking, social administration, etc., which is why the present paper is restricted to industry. Only a small fraction of workers have been employed to use such technology; these "islands" of advanced technology and the people who work on them usually enjoy special attention and treatment (greater prestige, training, incentives, etc.); the workers have been recruited on a voluntary or merit basis from the ranks of a growing workforce. (See table 1).

(c) The workshops and plants which make use of advanced technology have often proved to be vulnerable to deficiencies in management (problems of maintenance, supply of tools, materials, components, etc.) which have changed both the content and intensity of work for the workers as well as their efficiency. Such problems have emerged in Hungary, Poland and in the Soviet Union (see, for example, Zemlansky et al., 1977). When the organisation of autonomous "brigades" (which will be discussed later) was initiated some years ago by a worker called Zlobin in the Soviet construction industry and also introduced in a few other enterprises, including the Volga Automobile Works, Hatchaturov (1977) remarked: "Zlobin's method is an excellent example, but the fact that today only 1.5 to 2 per cent of the brigades follow this method highlights the difficulties of this new type of work organisation, of the supply of materials and machines".

(d) Physical strain rather than mental stress seems to cause more problems for workers in the centrally planned economies even today. Data from five countries shows that workers consider their jobs physically more strenuous than average and in some countries up to 25 to 33 per cent of unskilled workers and 20 to 25 per cent of skilled workers look upon their jobs as physically very or extremely strenuous (Akszentievics, 1982). (See Table 2)

Table 1: The composition of the manual work force according to the technological characteristics (mechanisation) of their work in the state-owned industry in Hungary (%) (1979)

	Supervisory controlling work	Simple manual work	Craftsman type work	in "line" organisation		at machines			Other	Alto-gether
				Manual work	Operator work	Manual work	Operator work	Supervisory controlling work		
Mining	3.0	19.3	23.7	1.5	0.7	17.7	21.5	5.4	7.2	100.0
Electric energy supply	5.6	19.0	33.4	0.8	0.6	5.6	14.1	9.7	11.2	100.0
Metallurgy	8.6	15.1	19.0	1.3	0.6	18.1	22.1	4.8	9.5	100.0
Machines and equipment	6.8	18.5	26.9	3.1	0.9	13.4	21.4	1.1	7.9	100.0
Vehicles industry	6.6	23.3	28.2	4.5	2.1	8.5	16.6	1.6	6.6	100.0
Electric machines and equipment industry	9.3	20.1	18.9	9.4	3.0	11.3	19.3	1.8	6.9	100.0
Telecommunications and vacuum technics industry	8.7	18.6	19.2	12.5	4.2	11.5	14.8	3.5	7.0	100.0
Instruments industry	9.4	17.6	26.5	10.1	2.8	7.2	17.2	1.7	7.5	100.0
Metal mass products industry	6.9	20.6	14.5	5.7	3.0	13.1	25.1	4.8	6.3	100.0
Construction materials industry	4.6	24.3	14.6	7.0	3.1	15.8	17.4	5.7	7.5	100.0
Chemical industry	9.3	20.5	19.7	2.1	1.2	13.6	16.9	8.7	8.0	100.0
Wood manufacturing industry	3.7	28.6	11.6	1.9	6.9	14.2	16.7	2.1	5.3	100.0
Paper industry	8.9	23.3	11.2	0.5	0.4	20.5	24.8	4.7	5.7	100.0
Printing	6.5	21.5	17.7	0.8	0.6	20.0	24.1	2.2	6.6	100.0
Textile industry	6.9	20.8	10.3	3.9	6.6	12.2	29.0	2.0	8.3	100.0
Leather, fur, shoe industry	5.4	13.5	7.9	20.6	27.8	6.1	12.2	1.5	5.0	100.0
Textile clothing industry	5.9	10.6	8.4	10.2	53.6	2.3	3.9	0.5	4.6	100.0
Handcraft and home industry	9.2	22.8	9.5	8.5	22.3	8.3	10.6	0.5	3.3	100.0
Food industry	5.9	30.0	11.2	9.4	2.9	11.4	14.2	4.5	10.5	100.0
State-owned industry	6.7	21.8	18.3	6.1	4.1	12.5	18.5	3.9	8.1	100.0

Source: Héthy et al., 1982.

Table 2: Physical and mental strain of jobs in five centrally planned economies as perceived by the workers (1979)

	Physical strain				Mental strain			
	Average level (on a 9-point scale)		Those in very or extremely strenuous jobs (%)		Average level (on a 9-point scale)		Those in very or extremely strenuous jobs (%)	
	un-skilled	skilled	un-skilled	skilled	un-skilled	skilled	un-skilled	skilled
Bulgaria	5.7	5.4	33.6	25.9	4.2	5.1	12	31
Czechoslovakia	5.4	5.1	27.6	19.2	4.5	4.7	29	19
Poland	5.5	5.0	21.9	14.3	4.5	4.9	22	26
Hungary	5.0	4.9	18.5	12.2	3.7	4.5	10	15
GDR	4.8	4.4	9.9	5.0	3.5	4.7	11	21

Source: Akszentievics, 1982.

(e) The introduction of advanced technologies has until recently mainly affected new entrants to the labour force and has not been accompanied by retraining of the existing labour force. As a result, workers with traditional skills have rarely been subject to negative changes such as obsolescence of their skills, etc., because they are relatively scarce. Their skills were also badly needed in the growing number of conventional plants and if they were attracted by new technologies, they were usually employed as set-up or maintenance personnel.

(f) A large number of the workers have been recruited either from among the peasantry or from among previously non-working women who had no industrial traditions and were less sensitive to the negative changes caused by technological progress in their conditions of work. These people had no past experience which could allow them to compare the "splintering of work tasks" unlike their skilled counterparts and, therefore, they did not look upon the introduction of new methods as a "narrowing" of their activities or of their autonomy, but as a challenge to cope with the requirements of industry, to achieve a level and degree of precision and attention, to endure a level of mental strain and to meet a discipline that had been unknown in their agricultural or household work. Thus, more advanced technology, even in its most notorious forms such as assemly lines, have for the most part been functioning as a training "school" to help workers become accustomed to the requirements of industry (Afanasiev, 1973). Although certain types of negative behaviour by the workers (turnover, reduction in performance, idleness, violation of work discipline and technological discipline, etc.) are present in the centrally planned economies, they are only partly related to dissatisfaction with job content.

(g) Given that industrialisation has developed simultaneously with a very remarkable increase in the skills and general level of education of workers, present day and future conflicts between the expectations of an educated workforce and the requirements set by the jobs and up-to-date technology cannot be ruled out.

(h) In the centrally planned economies, industrialisation (i.e. technological progress) has taken place at the same time as a radical, almost "revolutionary", improvement in the living standards of workers; as a result, it has appeared to them, justifiably, as a means of improving their lives. Considering the miserable living conditions of the workers in the past and the present level of their living standards in most eastern European countries, it is no wonder that many workers (especially those of peasant origin) are now aware of the material consequences of technological progress (that is, to its impact on their wages) than to its other possible concomitants concerning, for example, the content of their job. This is widely felt in Hungary (Héthy-Makó, 1981; Héthy et al., 1982).

3. In the 1960s and 1970s, the above features of the social-economic context of technological progress in the centrally planned economies hindered rather than promoted work content, work intensity and job satisfaction becoming a topical problem of industrial relations. Today, the influence of these factors and certain new developments such as increased level of education, skills, living standards of workers, growing industrial traditions, the draining and lack of further labour reserves, the slowing down or even stagnation of industrial growth and the pressing need for industrial efficiency and flexibility might easily bring the problems of work content, work intensity and job satisfaction into the focus of attention in the 1980s.

II. Automation, work organisation, work content and work intensity

4. It is not easy to describe the impact of automation and work organisation on job content, work intensity and above all on the workers because both the independent and dependent variables in this relationship are hard to define. Firstly, automation and work organisation, although closely related to each other, are two separate variables (for example, automated machines in engineering can be used as part of a large-scale, mass, and, at least in principle, single unit production of traditional non-flow or rigid flow production). Secondly, automation itself has differing stages and forms which produce very different effects on job content and work intensity (for example, numerically controlled machines in engineering or remote or computer controlled machinery in the chemical industry). Thirdly, job content and work intensity are not the same; both of them can be interpreted in many ways according to which factors (monotony, variety, creativity, responsibility, mental and physical strain, etc.) are given greater or lesser emphasis. Finally, the relationships between job content, work intensity and the workers in many respects are subjective in the nature, depending on the given needs and expectations of workers which originate from their social-economic environment. Monotonous, boring jobs are often more attractive to workers than creative, challenging ones, despite the expectations of social

scientists. Which is why the analysis of this question in this paper tries to start out from a differentiated approach of technology and work organisation, job content and workers' expectations concerning their work.

A. Alternative concepts of technology

5. Technological progress has been analysed in innumerable ways for the past 30 years. However, the approaches (Héthy-Makó, 1981) can be classified into two major groups. The first is the narrower approach of technical sciences which concentrates on the level of the mechanisation and automation of the direct and indirect production activities and on the mental and physical requirements of the job originating from them, a characteristic and well-known example of which is the technological classification by Bright (1958) or Auerhan (1961). The latter constructed an eleven-point scale for the measurement of technological development, placing manual work and work with hand-operated tools at one end of his scale, while the other extreme, which represents the most developed technology, is automated machinery which "solves not only the technical, but also the economic control of the production process". The second one is the broader approach of the social sciences, especially of sociology, that try to grasp all those characteristics of technological progress that influence the overall conditions of jobs, including its social aspects. Woodward's (1965) "technical complexity scale", constructed for the purpose of her research in the field of organisations, is a good example of this approach: in her description, single unit or small batch production to customers' individual demands represents the oldest and simplest way, while the most developed type is automated continuous-flow production process of so-called dimensional goods (chemicals, etc.). Touraine (1970), who also follows the historical trend of technological progress from the very start of the transformation of industrial work, makes a distinction among the professional, transitory and technical systems of production. In this latter approach, the role of work organisation beyond and even prior to mechanisation and automation, as an attribute of single unit, batch and mass production, flow and non-flow production, is emphasised.

6. Technological development today rarely appears as an unbroken trend from workshop type single unit or batch production on conventional machines to continuous flow production on automated machinery in engineering mass production or the chemical industry, but rather as a multilateral process in which the growth of modern technology is

accompanied by a kind of renaissance of the most traditional almost craftsmanship-type single unit or small batch production, in small enterprises. Consequently, it is a vexing question of what to compare: job content in automated production, but in what type of automation with what type of other technology as an earlier stage? How should work organisation be treated in this comparison? How should the differentiated needs and expectations by workers be taken into consideration?

B. Education and training

7. Summing up research findings in eastern Europe in this field (Auerhan, 1965; Krevnevich, 1971; Szteriosz-Denke, 1979; etc.), two conclusions seem to be obvious: firstly, technological progress in general and also automation are accompanied by an increasing level of the general education and qualifications of the workforce; secondly, it is unlikely that this parallel trend in automation and the level of education and qualifications of workers would lead to such an extreme situation in which traditional unskilled and semi-skilled jobs disappeared or physical work was substituted by mental work. It is more likely that we will witness a type of restructuring of the education and qualifications of workers, limited by the natural economic and social boundaries of automation. It should also be noted that the general level of education and qualifications of workers is only partly dependent on technological progress - education today is an autonomous process governed by autonomous social values.

8. During the past few decades, many workers have been affected by the transition from universal, multi-purpose machines to semi-automated machinery and mechanised lines. In this process (Auerhan, 1965; Szeteriosz-Denke, 1979), the proportion of unskilled work seemed to decrease; it dropped from 60 per cent to 20 per cent and then rose again up to 33 per cent, while semi-skilled work increased from 20 per cent up to 65 per cent and finally stabilised at 57 per cent. At the same time, the ratio of workers with middle-level and high-level education doubled from four to eight and from one to two per cent, but stayed at a fairly low level. (Data for the Soviet Union are higher, but this can be attributed, at least partly, to the intricacies of statistics.) In this respect, it seems to be obvious that the level of education and qualifications of workers has not been subject to radical changes.

9. When automation is taken into consideration, developments, although they affect only a small number of workers, seem to be much more radical. According to Auerhan, 1965; and Szteriosz-Denke, 1979, only 12.5 per cent and 4 per cent of the workers respectively have middle and higher-level education; however, their proportion in case of fully automated technological processes amounts to 60 per cent and 34 per cent respectively and unskilled labour practically disappears. It should be noted, however, that the number of workers employed in fully automated technological processes tends to be insignificant in relation to the whole of the workforce and in this way the problems of these people, although they might be important for the future, do not seem to be pressing for the present. (See Table 3)

10. Auerhan's forecasts concerning the structure of the qualifications of the labour force which vary according to the stage of technological development realistically reflect the general trends, but are rather abstract. In industrial conditions it seems somewhat illusionist to believe in "islands" of up-to-date technology not having any roots in supporting sectors populated by semi-skilled or unskilled labour. The presence of unskilled workers in these sectors, as described by Szczepanski in 1977 (to be quoted later on) is a natural concomitant of technological progress in eastern Europe.

C. The importance of work organisation as related to automation

11. The contents, requirements and conditions of jobs seem to be influenced by the characteristics of work organisation to an even greater extent than they are by the level of technological development. This is discussed below.

12. A very important aspect of work organisation seems to be the volume of production: in this respect, distinctions are usually made between single-unit, small-batch, large-batch and mass production. The most significant difference between the two extremes is that single-unit production is carried out at individual customer's demands, according to individual specifications and consequently the work, at least in principle, is not repeated. Mass production, on the other hand, embodies the manufacturing of standard goods under standardised conditions for longer periods of time (in the automotive industry, it is of the order of several months), thus tasks are constantly

Table 3: The qualification of labour in the stages of technological progress (%)

Qualification of labour	Stages of technological progress								
	3rd	4th	5th	6th	7th	8th	9th	10th	11th
	mechanisation			automation					
Unskilled	15	7	-	-	-	-	-	-	-
Semi-skilled	20	65	57	38	11	3	-	-	-
Skilled	60	20	33	45	60	55	40	21	-
Middle level ed.	4	6.5	8	12.5	21	30	40	50	60
High level ed.	1	1.5	2	4	7	10	17	25	34
Scientific degree	-	-	-	0.5	1	2	3	4	6

Stages of technological development: 3rd: universal machine; 4th: semi-automated machine; 5th: mechanised line; 6th: automated machine; 7th: automated equipment; 8th: self-controlled automated equipment; 9th: automated equipment with self-controlled registration of the main indices of production; 10th: automated equipment with self-controlled optimation; 11th: fully automated production equipment.

Source: Auerhan, 1965, quoted in Richta et al, 1968.

repeated. Mass production and standardisation have made it possible for the establishment of types of work organisation characterised by a high level of specialisation. (The fragmentation of work, as is typical of assembly lines in the automotive industry, was used for the first time some seventy years ago by Ford.) The importance of the volume of production, compared to that of mechanisation and automation, from the point of view of the content of industrial work, should, therefore, be underlined.

13. Another important aspect of work organisation, at least in the engineering industry, is the pattern of the production process, that is, its non-flow or flow character. In the case of non-flow production, the machining is carried out on a large number of pieces, operation by operation; that is, the organisation of the work process is concentrated on individual operations and machines. In a workshop organised according to this traditional pattern, the machine tools are located in homogeneous groups, that is, lathes are put in one group, boring machines in another group, etc; thus the location of the machinery has nothing to do with the sequence of operations on a certain product. It often happens that pieces are subjected to operations by one machine only, or by a small number of the machines, and they often leave the workshop without having been worked on by most of the machines. In the case of flow production, on the other hand, organisation is focused on a set of operations and machines are assigned to the same product or to a small group of products. The location of the machinery more or less follows the sequence of operations on the piece. The product goes through most or all of the machines in the course of its manufacture. In this pattern of organisation, the individual machines are connected by transport equipment, as exemplified by the production lines in the automotive industry. Lines can be flexible or rigid. With the first system, stocks (buffers) of pieces can be established between two machines in the line, or the sequence of operations can be partly changed. In the second case, operations by the individual machines are closely linked: the second machine can only start working when the first machine has finished: no buffers can be established and no changes in the sequence of operations can be realised. A typical example of rigid flow production is an automated transfer line. It should be noted that assembly lines also represent flow production and they can be organised in both flexible and rigid ways.

14. To test the relative importance of the above features for the content and conditions of industrial work, an investigation of their impact on the Hungarian automotive industry (Héthy-Makó, 1981) was carried out. Jobs

and production units that differed in just one of these dimensions, while being strictly homogeneous in the other two dimensions, were compared. The analysis included a comparison of: (a) jobs on automated, semi-automated and conventional machines (all from large-batch and flow production), (b) jobs in single-unit and large-batch production (all from conventional machines and in non-flow production), (c) jobs in traditional non-flow and flow production (all from large-batch production on conventional machines). This analysis, which did not cover all the possible combinations of the three dimensions, has produced the results shown in Tables 4 to 8.

15. Of all the aspects of technological progress that were examined, it was the progress from single unit production in the direction of large-batch production that affected the operators' work content most. There is a world of difference between a worker working on the same conventional machine tool in single-unit and in large-batch production. In regard to their work content, large-batch operators' jobs are considered narrower in scope and "poorer" than jobs in single-unit production.

16. Progress from traditional non-flow production towards flow production has an inconsistent effect on work. On the one hand, flow production widens the tasks and lends them variety (for example, machine tool operation expands to cover the operation of transport and hoisting equipment). But, on the other hand, it reduces independence and adds to the restrictions on the workers.

17. Progress from conventional machines to semi-automated and automated machines brings about no essential change in work content, even if work on an automated machine appears to be richer in content because of its novelty.

18. Regarding working conditions, little relationship was found with the aspects of technological progress examined. On the other hand, it has been proved beyond doubt that single-unit production was the most attractive form of work for the workers regarding both wages and the job's physical conditions. Data also indicated that the development of working conditions is related to the nature of the prevalent technology. For instance, flow production is usually accompanied by a collective incentive system and non-flow production by individual incentives. In fact, physical working conditions in large-batch production on production lines developed in entirely different ways in

Table 4: Work content as perceived by workers in large batch and single unit production

Job characteristics (present to a "large extent"*)	Traditional non-flow production on conventional machines			
	batch (n = 110)		batch (n = 99)	
	%	rank	%	rank
1. Variety	33.6	(7)	92.9	(1)
2. Independence	49.1	(3)	87.9	(3)
3. Responsibility	73.7	(1)	88.9	(2)
4. Chance of using one's knowledge and training	47.3	(4)	80.9	(4)
5. Possibility of working out new and better ways to do one's job	17.3	(8)	13.2	(8)
6. The need to learn new things	39.1	(5)	53.6	(7)
7. Interesting work	67.2	(2)	79.8	(5)
8. Chance of developing one's own abilities	38.2	(6)	63.7	(6)

* "To a large extent" corresponds to points 7-9 on a 9-point scale

Source: Héthy-Makó, 1981.

Table 5: Work force structure in large batch and single unit production

Worker characteristics	Traditional non-flow production with conventional machines	
	single unit production - % (n = 99)	large batch production - % (n = 110)
1. Men	81.8	26.4
Women	18.2	73.6
2. Workers with elementary school certificates and skilled workers' diplomas	82.8	32.8
3. Skilled workers	87.9	37.3
4. Seniority of more than 10 years with the job	50.5	13.7

Source: Héthy-Makó, 1981.

Table 6: Work content as perceived by workers on automated and conventional machines

Job characteristics (present to a "large extent"*)	Large batch flow production			
	operators of			
	automated machines (n = 48)		conventional machines (n = 115)	
	%	rank	%	rank
1. Variety	44.9	(3)	28.7	(5)
2. Independence	42.9	(4)	67.8	(2)
3. Responsibility	79.6	(1)	80.0	(1)
4. Chance of using one's knowledge and training	32.6	(7)	37.4	(4)
5. Possibility of working out new and better ways of doing one's job	10.3	(8)	18.3	(7)
6. The need to learn new things	38.8	(5)	17.4	(8)
7. Interesting work	57.1	(2)	38.2	(3)
8. Chance of developing one's own abilities	32.7	(6)	20.9	(6)

* "To a large extent" corresponds to points 7-9 on a 9-point scale.

Source: Héthy-Makó, 1981.

Table 7: Work content as perceived by operators and non-operators

Job characteristics (present to a "large extent"*)	Automated				Non-automated			
	operators		non-operators		operators		non-operators	
	% (n = 40)	rank	% (n = 62)	rank	% (n = 288)	rank	% (n = 53)	rank
1. Variety	44.9	(3)	80.5	(1)	44.9	(5)	69.7	(3)
2. Independence	42.9	(4)	57.4	(5)	66.9	(2)	54.7	(7)
3. Responsibility	79.6	(1)	78.7	(2)	80.9	(1)	71.6	(2)
4. Chance of using one's knowledge and training	32.6	(7)	67.2	(3)	50.0	(4)	66.0	(4)
5. Possibility of working out new and better ways to do one's job	10.3	(8)	26.2	(8)	18.8	(8)	32.1	(8)
6. The need to learn new things	38.8	(5)	52.4	(7)	30.2	(7)	62.3	(5)
7. Interesting work	57.1	(2)	60.6	(4)	56.3	(3)	79.2	(1)
8. Chance of developing one's own abilities	32.7	(6)	54.1	(6)	37.1	(6)	58.4	(6)

* "To a large extent" corresponds to points 7-9 on a 9-point scale.

Source: Héthy-Makó, 1981.

Table 8: Work force structure of operators and non-operators

Worker characteristics	On automated machines		On conventional machines	
	operators % (n = 40)	non-operators % (n = 62)	operators % (n = 288)	non-operators % (n = 53)
1. Men	91.8	98.4	60.1	98.1
Women	8.2	1.6	39.9	1.9
2. Technician	8.4	14.8	2.1	1.9
Skilled worker	56.2	85.2	51.4	96.2
Semi-skilled worker	35.4	--	46.5	1.9
3. Seniority of more than 10 years with job	32.6	67.2	33.0	47.1

Source: Héthy-Makó, 1981.

the plants, as did the incentive systems. The development of working conditions depends on management policies rather than on technological progress.

D. Partial automation, job content and working conditions

19. Comparative empirical research on changes caused by automation in job content, work intensity and working conditions in the centrally planned economies has produced very few conclusive results. Most of them appeared in the "Automation and industrial workers" project (Forslin et al., 1981). The findings on physical and mental strain were that certain types of automated techniques in the automotive industry (transfer lines) lead to a probable decrease in physical strain and to an increase in mental strain for the workers when compared to certain traditional technologies (large-scale or mass-production on conventional or semi-automated machinery).

20. Such findings were described in the Soviet report for this project (Ussenin et al., 1979) as follows:

> "Manual operations occupy almost all of the time of non-automated jobs with the bulk, up to 55 per cent, on loading, fixing and unloading the material and 7 per cent for immediate control of the work process ... On the contrary, workers in automated units, as a rule, spend not more than one per cent of their time on manual operations. The operators of these units spend 5 per cent of their work time on loading, fixing and unloading ... This has had a forcible influence on the changing ratio between the workers' physical and mental efforts in the work process. The general score of all functions related to the mental operations of the workers of automated production appears higher than that at non-automated units, while the share of physical efforts is reduced. The level of physical strain connected with lifting, carrying, pulling, pushing, retrieval, etc., of various loads was considerably reduced ... The numerical value of the resulting index of physical strain for those items produced in automated production is 5.2 verus 11.3 in non-automated production."

The German Democratic Republic report speaks about "reducing heavy labour through the introduction of automation" and states: "Transition to automation is accompanied by changes in the relationship between manual and intellectual labour" (Adler et al., 1981). The Polish, Czechoslovak and Hungarian findings are, however, rather vague concerning the balance of physical and mental strain caused by automation. The Polish report says: "Although workplace observations show that both mental and physical strain is likely to decrease as a result of automation, non-automated workers perceived less mental strain than automated ones" (Sarapata, 1979). The Czechoslovak report states that: "The work-load and physical demands of work showed no statistical difference. In non-automated units, however, there seems to be greater emphasis on improvement of both of these problems" (Charvat et al., 1981). The Hungarian report states that: "Although workplace observations have revealed some minor differences in the mental and physical requirements of automated, semi-automated and conventional machine operators, these differences do not fall into a clear-cut and unambiguous pattern" (Héthy-Makó, 1981).

21. None the less, automation seems to enrich job content in other respects, at least as it is perceived by the workers. The Soviet report for this project (Ussenin et al., 1979) records a slight but not insignificant improvement in variety, independence, responsibility, opportunities to use knowledge and training, possibilities of working out new and better ways to do the job, the need to learn new things, interesting work and the chance to develop abilities. Similar results were borne out in the German Democratic Republic and in Hungary, although in the latter case certain elements of job content showed slight deterioration as a result of automation: for example, loss of independence (Adler et al., 1981; Héthy-Makó, 1981). In Hungary, workers in automated units perceived wider possibilities to work out new and better ways of doing their job, to learn new things and to develop their abilities, but generally they considered such opportunities as being rather restricted (Héthy-Makó, 1981). In Poland (Sarapata, 1979), workers perceived wider possibilities for promotion in automated units than in conventional ones - an opinion shared by employees in some other countries. In Czechoslovakia (Charvat et al., 1981), automated jobs were found much more collective-oriented than others.

22. As for work intensity and working conditions, the research findings are more confused. The Soviet report states: "... the general working conditions connected with equipment and confort of physical environment improve substantially ... Automation leads to a general reduction in work intensity and accordingly lessens worker fatigue". It lists improvements in air contamination, cleanliness, humidity and space for the workers (Ussenin et al., 1979). At the same time, reports from other countries arrive at more sceptical conclusions. The Ezechoslovak paper indicates a "higher probability of accidents and danger of accidents as well as poorer physical conditions" in the automated units (Charvat et al., 1981). The Hungarian paper concludes, in a comparision of automated and non-automated jobs, that: "In general, there were many complaints about the physical working conditions, including air contamination, noise, inadequate illumination, dirty environment and time pressure ... Differences in working conditions within the various groups - for instance, between traditional machine jobs - are often significantly more important than differences between automated, semi-automated and conventional jobs " (Héthy-Makó, 1981).

23. Previous investigations in the Hungarian machine and steel industry concluded that work intensity did increase as a result of the introduction of new technology. The work intensity of different technological development stages (classified according to Touraine's categories) was investigated in Hungary in the early 1970s and it was found that workers generally perceived that it increased as a result of the introduction of new technology both in the machine and steel industry (Héthy-Makó, 1976, quoted by Szteriosz-Denke, 1979). In machine industry plants, 49.4 per cent of the workers reported that work intensity had increased compared to the requirements of the previous technology, while 13.3 per cent perceived a decrease. In steel industry plants, the corresponding proportions were 45.5 and 4 per cent.

24. How can these contradictions be explained? The transfer-lines (and NC machines) investigated in the five centrally planned economies represented only a partial level of automation[3] with many manual operations still required. This equipment, which was frequently introduced as a herald of new technology, was often connected up to much less developed machinery and used as part of a "mixed unit", in "inadequate premises" - as the Czechoslovak report states - which essentially determined the physical conditions for their workers. As investment funds were scarce, enterprises frequently economised by only purchasing the productive part of the technology and not

the support equipment which would decrease physical strain (such as hoisting and lifting equipment) or improve the physical environment - ventilation, air-conditioning, etc. In addition, these machines were very often used to the limit to ease bottlenecks and they were exploited by management to a great extent - which also contributed to the physical and mental strain of those working on them. The influence of these factors was particularly felt, it should be emphasised, when conventional plants were widened by "islands" of new technology, as has occurred often in the past and which is also very likely to happen in the future.

E. Problems of advanced automation

25. More advanced forms of automation have been introduced in branches using continuous flow production such as metallurgy, oil refining, paper mills, power stations, chemicals,etc. In such cases, part or all of the production process is automated. Intervention by the operator is only necessary in certain phases, or in case of emergency, when he may be assisted by a computer. Although we have no comparative data about the effects of this phase of technological progress in the centrally planned economies, the sparse information at our disposal makes it possible to underline the presence of certain new implications (see, for example, Andics, 1978; Antalovits, 1977; Richta, 1968; Szteriosz-Denke, 1979, etc.):

(a) Physical strain for the workers becomes virtually non-existent.

(b) Monotony, the most crucial problem of automated large-scale or mass production in engineering, seems to be substituted by vigilance. Monotony only continues to exist in cases where an important phase of the production process still needs constant manual control by the operator, but the task is not challenging enough to make use of the worker's mental capacities. In other cases, the need for vigilance becomes even more emphasised. In the computer-controlled production process of a paper mill recently opened in Hungary, the operator's job requires either quick intervention on the basis of a narrow predetermined set of reactions or a cognitive decision based on partly unknown reactions selected with the aid of a computer.

(c) Responibility appears to take on a new meaning in jobs with multiple requirements and sometimes its nature becomes rather vague and obscure. A responsible operator in such a situation is not the one who works industriously and keeps to the predetermined rules of technology and discipline, but who keeps himself in good physical and mental condition so as to intervene quickly in an emergency and who is able to function with little supervisory control.

(d) Social isolation on automated production lines, due to noise and work intensity, makes communication difficult or impossible. In the case of remote or computer-controlled processes, operators are often physically isolated (locked up in their cabins with their instruments), though this does not necessarily lead to their social isolation. The control of such production processes depends not only on the contribution by one or other of the operators, but on their strictly co-ordinated collective action mediated by technical instruments (Zinchenko-Munipov, 1978). Thus, despite physical isolation, the success of production is based on "personal contacts".

(e) Physical conditions of work improve considerably.

In a comparison of traditional and continuous casting processes in the Hungarian steel industry, workers on the automated system perceived their jobs as being both physically and mentally less strenuous. They considered the likelihood of job-related accidents and illnesses much less menacing than did their colleagues involved in conventional production and did not complain of any overload of work (Makó, 1981). The explanation lies partly in the fact that machinery took over the very tiring manual and mental operations and certain hard, dirty jobs such as cleaning the moulds ceased to exist and partly because of the separation of the operations of the workers from the production process itself.

26. A radical improvement in the physical conditions of work appears to be a frequent concomitant of technological progress. This usually takes place when the operator works in a control room which is separated from the production machinery. However, this does not occur in the case of NC machines (Andics, 1977).

27. Thus, to sum up: "Looking upon the activities of the operators of the newly installed machines and lines, it often appears as if mechanisation and automation has made work easier. This impression, however, is not justified. In many cases, it is not so much the intensity of the strain that has changed, but rather that its nature has been modified. As a result of technological progress, information processes in the human beings (perception of

signals, taking decisions, executing decisions, i.e. transferring a decision to the machines) receive more emphasis; the importance of vigilance increases and responsibility for the consequences of possible mistakes and negligence becomes more acute. In short, the psychological strain on men increases, while the physical strain, in most cases, ceases to exist" (Antalovits, 1977).

F. Restructuring of operators' and non-operators' jobs

28. Technological progress brings about a restructuring of workers' jobs. This development is summarised by Szczepanski (1977) on the basis of considerable sociological research in Poland, as follows:

> " ... technological development and advancing automation brings about changes in the very content of work, in the nature of the activities (operations) by the workers, and at the same time it also modifies the ratio of mental work in these activities. Workers, on the basis of the content of their work, can be put into several categories, such as: (a) simple manual workers doing unskilled work, (b) semi-skilled workers using machines, (c) skilled workers using machines, setting up the machines and carrying out various operations on them, (d) set-up men, (e) workers controlling automated machinery, (f) workers engaged in the repair and maintenance of machines, including automated machinery ... Experience in both capitalist and socialist countries of automated plants has proved that these establishments always need workers in category 'a' to do simple manual work. At the same time, automated factories employ a large number of work groups engaged in repairing and maintenance, whose members are highly qualified but often engaged in physically very strenuous work. In categories such as 'd' and 'f', the work required such a high level of qualification that the use of the workers' knowledge is already intertwined by creative elements."

29. Two points of Szczepanski's evaluation, which constitutes another possible social science approach to automation, deserve more emphasis. Firstly, it is unlikely that technological progress in our time will eliminate the most unpleasant jobs, those that are physically hard and

carried out in unfavourable conditions. Secondly, that automation in any form, including that applied to assembly lines, leaves certain skilled set-up and maintenance jobs untouched and may even add to their importance. These jobs are "wider, richer and have more substance than any operator's job except perhaps for those in single unit production" (Héthy-Makó, 1981). This phenomenon indicates that even in the most rigid forms of standardised mass production which leads to the fragmenting of work tasks, there are possibilities for enriching job content.

G. Workers' needs and attitudes towards technological progress

30. A reliable knowledge of the types of workers' needs is of strategic importance in formulating policies concerning the development of both technology and work organisation, in order to keep control of the social consequences of these processes. However, social research has produced very little information about them for the past 15 to 20 years in the centrally planned economies, except for a few admirable but rather isolated examples of which the study by Zdravomyslov et al. (1978) is one. The results that are at our disposal offer some rather vague conclusions:

(a) Work content seems to have enough importance to merit the attention of both social science and social administration. "The content of the job, the creative possibilities of work, are the specific factors which determine a worker's attitude to work" (Zdravomyslov et al., 1970). This formulation is also more or less supported by the findings of the "Automation and Industrial Workers" project (Ussenin et al., 1979; Sarapata, 1979; Héthy-Makó, 1981; Makó, 1981).

(b) A job that is rich in content is only one of the expectations of workers which is often over-shadowed or thrust into the background by other equally or more important needs, at least in some countries. The Hungarian report speaks about the "predominance" of material needs of the workers, stating: "The introduction of new machinery, equipment and production procedures emerges for the workers primarily as a wage issue to be solved as such by the company" (Héthy-Makó, 1981). Although Soviet workers ranked "good pay" rather low, "upgrading and promotion" was given a high score by them and "upgrading or promotion ultimately ensures higher pay" (Ussenin et al., 1979). This phenomenon is rooted, in our opinion, in the level of the satisfaction of basic material needs, food, clothing, housing and durable consumer goods, such as furniture, household machinery, automobiles; in the supply of such goods in the consumer market; in the housing shortage, and other living standards, in the workers' way of living and in the social and ideological values, etc. In these conditions there is much similarity, but also many differences among the centrally planned economies.

(c) The structure of the workers' felt needs and expectations concerning their work is very differentiated even within the same country according to the level of education, skills, sex, age, etc., of the workers. They cannot be looked upon as a homogeneous group, even if they are motivated by those needs that seem to be predominant, e.g. to have a creative job or good pay (Héthy-Makó, 1981).

31. Workers in the centrally planned economies seem to cherish a definite positive feeling towards the introduction of new technology. In the "Automation and Industrial Workers" project, a vast majority of the workers in the five countries thought that "the introduction of new machinery favourably affected the workers" and declared that "if labour-saving new machines were introduced in their plant they would either actively support or quietly approve such changes". The proportion of workers endorsing these positive answers were as follows: in Czechoslovakia 86.7 per cent, and 100 per cent; in Poland 70.5 per cent and 97 per cent; in Hungary 74.8 per cent and 89.7 per cent; in the German Democratic Republic 80.7 per cent and 99.4 per cent; and in the Soviet Union 98.6 per cent and 100 per cent (Makó, 1981). This attitude, which can be explained by the whole historical, social and economic context of technological progress in these countries, is the more remarkable since the impact of automation in the automotive industry in these countries on work intensity and physical conditions of work was far from unambiguously positive. Job security is probably the main factor in this attitude.

III. New forms of work organisation and technological improvements

32. Efforts to develop work organisation in the ideology of most centrally planned economies as formulated in the programmes of scientific work organisation have three major elements:[4] firstly, to "ensure the most effective use of labour and material resources and a steady increase in labour productivity", secondly, to contribute to "favourable working conditions and protection of the health of the people", and finally, to promote "the general development of workers" and "to transform work itself into the most meaningful human activity" (Dovba, et al., 1979; Hanspach-Schäfer, 1979).

33. The new forms of work organisation in the centrally planned economies follow in their methods those new and old paths that are generally considered well-known (see, for example, Miller and Form, 1964; Delamotte, 1972; etc.). There is an effort being made in each country to eliminate physically or mentally strenuous, hard and dirty work by mechanisation and automation (see, for example, Hanspach-Schäfer, 1979; Kalinina-Roik, 1983). To our knowledge, however, these efforts have not reached the stage where complete technological processes and work organisation patterns are reformed to meet the workers' expectations.[5] Workers are selected "for particularly monotonous and demanding jobs" on the basis of "medico-biological and psychological criteria of suitability" (Dovba et al., 1979) as well as according to the "demands of workers" in which "jobs rich in content are allotted to people who most require such jobs and vice versa". In this way, monotonous, restrictive semi-skilled jobs often give the same amount of job satisfaction as more interesting, creative, autonomous jobs (Héthy-Makó, 1981).

34. Because of the money-oriented attitude of most workers, it is possible in some countries, and particularly in Hungary , to compensate workers with extra money for certain negative aspects of industrial work (physical and mental strain, unpleasant physical working conditions, etc.). In Hungary, such factors are built into the construction of the national job evaluation and wage categorisation system.

35. Job enlargement (the increase in the number of tasks performed by the employee in his/her work) and job enrichment (the addition of tasks of differing quality, i.e. set-up, maintenance and supervisory tasks) have been widely used in most of these countries. In the German Democratic Republic, where these methods are especially

widespread in the electronics industry, a distinction is made between "merging operations, especially those that have to be performed on each workpiece" and "the merger of the two kinds of tasks", i.e. operator and auxiliary work (Hanspach-Schäfer, 1979), but often such solutions are simply instinctively adopted by management (Héthy-Makó, 1981).

36. Job enlargement and enrichment in engineering and automotive industry is often realised in close connection with multi-machine operations and group work which have been given considerable emphasis, especially in the Soviet Union. "A current trend in industry is to combine the development of multi-machine operation and group-work organisation: the number of machines operated by a team is considerably greater than the number of team members. In comparison with individual work organisation, organisation of work in teams leads to a considerable reduction in the amount of equipment standing idle pending maintenance. In addition, there are more opportunities for increasing the workers' skills." Multi-machine operations are based on multi-skilling. "In many Soviet undertakings, 20 to 50 per cent of all workers are multi-skilled" (Dovba et al., 1979).

37. Job rotation (the movement of workers among different jobs) occurs in two major ways: firstly, when workers engaged in a definite phase of production systematically exchange their (mostly operator) jobs; secondly, when production workers are moved, from time to time, to auxiliary activities such as maintenance, preparation, etc., for retraining. The first one is closely related to group work and job enlargement, the second one to job enrichment. Both are connected with multi-skilling (Hanspach-Schäfer, 1979; Dovba et al., 1979).

38. "Job rotation is a feature of team work, under which a team of workers is called upon to perform a complex assignment consisting of a series of different elements. The teams comprise between four and eight workers, who can perform all the operations involved and who move from one work station to another. The team is responsible for the entire assignment, including quality control. Conditions are the same for all its members, with regard, for example, to performance indices, wage scales and methods of payment", according to a report from the German Democratic Republic (Hanspach-Schäfer, 1979).

39. Among recent developments in the field of work groups the rapid growth of "brigades" in the Soviet Union since 1979-80 should be mentioned: in a well-organised and controlled process by the enterprises and the central

agencies,[6] work "brigades" have been established to carry out their activities in distinct, separate phases of production, in order to make the best use of machines, materials and their own creative energies. "Brigades", functioning in all branches of the economy, are specialised (with their members having homogeneous skills) or complex (members having different skills). Some brigades, in chemicals, metallurgy, mining, etc., are organised on an inter-shift basis and called "cross-over brigades". The collective efforts of these work groups are also motivated by collective money incentives. Brigades have a say in the division of premiums and in fixing the job categories for their members, etc. (Bogdan, 1983). It is obvious that the jobs of the members of such "brigades" are also enriched by participating in decisions orginally belonging to the first-line supervisors (foremen, senior foremen). The development of "brigades" in the Soviet economy has been based on experiences of individual enterprises prior to 1979, including, for example, the Volga Automobile Works (Dovba et al., 1979).

40. In Hungary, although the development of "brigades" is much less co-ordinated and controlled by the central agencies, similar trends have occurred since the 1960s: several enterprises (in engineering, in the machine industry) on their own initiative have given the right to certain work groups to divide their work tasks as well as the money they earn amongst themselves (Héthy-Makó, 1977).

41. The solutions described above are often accompanied by measures of an ergonomic nature to improve working conditions: changes in the speed of assembly lines (Dovba et al., 1979), introduction of work breaks, etc. Ergonomic measures are also applied to cope with problems raised by advanced automation.

42. As to the economic benefits of such efforts, existing data (see, for example, Dovba et al., 1979) seem to be exaggerated, while costs are unknown.

IV. The use of social mechanisms to influence the consequences of automation and work organisation

43. An institutional framework of participation in the management of the enterprises has been developed in all of the centrally planned economies in the course of the past decades: it includes possibilities for direct participation that might add to the widening of the workers' activities at the plant and thus compensate the possible narrowness of their jobs, and indirect participation (via workers' representatives which are the trade unions) that might act to control the use of technological progress and its consequences at the enterprises (e.g. Héthy-Mako, 1977; Héthy, 1981).

44. Workers have often developed informal channels to regulate their own work performance and work intensity, both in collective and in individual ways regardless of the formal requirements formulated by the organisation or dictated by the given technology. Such prominent informal activities were recorded by social scientists in Poland in the 1960s (Dőktor, 1966) and in Hungary (Héthy-Mako, 1972; Fazekas, 1982). The relation of informal structures and control of technology seems to be rather ambiguous: technological progress can add to informal autonomy, but can undermine it as well (Simonyi, 1978).

45. Workers' affiliation with the "secondary economy" (Héthy, 1982) has also functioned as a means of compensation for dissatisfaction with the content and conditions of jobs and wages.

46. As to the role of enterprises and central agencies, two somewhat differing models have been functioning for the past 15-20 years. A centralised mechanism has been followed primarily in the Soviet Union and in the German Democratic Republic: in it the problem of work organisation is treated as a part of the Scientific Work Organisation programmes, in the realisation of which a major role is played by the central agencies and central plan: "The national economic plan firmly establishes the most important and effective measures to be adopted in all branches of the national economy. These measures include the following: (a) models for the organisation of work stations for the more common trades, for engineers and other technical staff and for non-manual workers; (b) norms and standards which set output rates for the workers in the economy as a whole and for particular

branches; (c) model plans for organisation of work units within a workshop; (d) organisation charts of new forms of work structure; (e) the introduction of multi-machine operation; (f) multi-skilling; and (g) the introduction of forms of organisation adapted to group production". (For further details, see Dovba et al, 1979; Hanspach-Schäfer, 1979).

47. In contrast to this model, a more decentralised approach has been adopted by Hungary since 1968: it has been left for the autonomous decision-making enterprises to work out their own organisational structure, including their work organisation, to choose a suitable incentive system, to determine the performance standards (i.e. piece rates) for their workers, etc. These measures were taken by the enterprises themselves, i.e. it was a central tenet in the new economic guidance system that enterprises, having a considerable freedom to use their production resources and motivated by profitability, would endeavour to find the best possible solutions. Central intervention in autonomous decision-making has usually taken an indirect course and has happened only if the special interests of the workers or general social interests (which sometimes contradict profitability) had to be safeguarded. Thus central agencies have strictly controlled the application of centrally set labour safety regulations and the level and differentials of wages, etc.

48. Trade unions in all of the countries with centrally planned economies enjoy considerable rights and have their own strategies to influence technological progress and the development of work organisation at workshop, enterprises and national level. Their tasks and responsibilities in this field are listed in a number of documents adopted by most of them since the second half of the 1970s. Such a resolution was adopted by the Confederation of Free German Trade Unions in 1975, by the Hungarian Trade Unions in 1978, etc. Legislation, including the labour codes, also gives support in various forms to initiatives taken by the central agencies, enterprises, trade unions and the employees themselves.

Notes:

[1] Director, Research Institute of Labour, Budapest, Hungary.

[2] In view of the shortage of reliable scientific knowledge and data, our paper does not pretend to be a systematic comparative analysis of the centrally planned economies,

but rather a survey of existing information in which some countries unfortunately had to be neglected while Hungary received more emphasis than others.

[3] More advanced automation in this sector is represented by the introduction of industrial robots which is in an initial stage in the centrally planned economies: the Soviet Union and the German Democratic Republic employ the most, about 4-5,000 each, but little is known as yet about their impact on industrial work.

[4] In German, Wissentschaftliche Arbeitsorganisation (WAO). In Russian, Nauchnaja organizacija truda (NOT). Note: it is not identical with the classical scientific management approach (Taylorism), although it incorporates several of its modernised elements.

[5] As in the case of the Volvo Kalmar Plant (Agurén et al, 1976).

[6] For the central guidelines on organisation of brigades, see: Metodicheskie osnovi briadnoj formi organizacii i sztimulirovania truda v promislennosti (Methodological grounds of the "brigade" form of work organisation and stimulation in industry), Moscow, NIIT, 1981.

References

Adler, Fisher, Kretzschmar, Lötsch, Winzer and Wittich: "Scientific and technological progress and the social activities of workers in the GDR", in J. Forslin et al. (eds): Automation and Industrial Workers: A Fifteen Nation Study (Oxford, Pergamon Press, 1981), Vol. 1, Part 2, pp. 199-232.

V.G. Afanasiev: Iránvitás, képzés, tudományos-technikai forradalom (Management training, scientific-technical revolution) (Budapest, Kossuth Könyvkiadó-Közgazdasági és Jogi Könyvkiadó, 1973), 420 p. Russian original: Moscow, 1972.

S. Agurén, R. Hansson and K.G. Karlsson: The Volvo Kalmar Plant: The Impact of New Design on Work Organization (Stockholm, The Rationalization Council, SAF-LO, 1975).

G. Akszentievics: "Az iparban dolgozók helyzetének összehasonlico vizsgarata öt europai szocialista országban" (A comparative analysis of industrial employees in five European socialist countries), in Társadalomtudományi Közlemények (Budapest, Magyar Szocialista Munkaspart, 1982), No. 2, pp. 26-49.

J. Andics: A technikai haladás társadalmi problémái a gazdasági szeivezetekben (Social problems of technological progress in economic organisations) (Budapest, Akadémiai Kiadó, 1977).

M. Antalovits: "Az operátori tevékenység pszichológiai sajátosságai" (Psychological features of the operator's activity), in Ergonómia (Budapest), 1977, Vol. 10, No. 4, pp. 177-181.

J. Auerhan: Die Automatisierung und ihre ökonomische Bedeutung (Berlin, Verlag Die Wirtschaft, 1961).

J. Auerhan: Technika, kvalifikáce, vzdeláni (Prague, 1965). Quoted in Szteriosz-Denke 1979 and Richta 1968.

R. Blauner: Alienation and Freedom: The Factory Worker and His Industry (Chicago, University of Chicago Press, 1964).

L.S. Bljahman: Proizvodstvennüj kollektiv (Moscow, Izdatielstvo politicheskoj literatüri, 1978).

J. Bogdán: "A brigádmunka uj utjai a Szovjetunióban" (New ways of brigade work in the Soviet Union), in

Társadalmi Szemle (Budapest, Magyar Szocialista Munkaspart), 1983, No. 4, pp. 60-64.

J.R. Bright: Automation and Management (Boston, Harvard University Press, 1958).

F. Charvat, J. Rehák and M. Tucek: "The worker in the process of changing technology", the Czechoslovakian Report, in J. Forslin et al. (eds): Automation and Industrial Workers: A Fifteen Nation Study (Oxford, Pergamon Press, 1981), Vol. 1, Part 2, pp. 23-57.

Y. Delamotte: Recherches en vue d'une organisation plus humaine au travail (Paris, La Documentation Française, 1972).

Dóktor, Kazimierz: "Le conformisme des travailleurs aux pieces", in Sociologie du Travail (Paris), 1977, No. 1. [Special edition entitled "La sociologie industrielle en Pologne"].

A.S. Dovba, V.L. Shagalov, I.I. Shapiro and A.F. Zubkova: "USSR", in ILO: New Forms of Work Organisation (Geneva, ILO, 1979), Vol. 2, pp. 79-111.

P. Dubois and C. Makó: "La conquête de l'Ouest: L'habillement hongrois à l'assaut du marché capitaliste", in Revue Française des Affaires Sociales (Paris, Ministère de la Santé et de la Securité Sociale), Oct.-Dec. 1982, pp. 87-99.

"Ember, tudomány, technika: Kisérlet a tudományos-technikai forradalom marxista elemzésére" (Man, science, technology: A contribution to the Marxist analysis of scientific-technical revolution) (Budapest, Kossuth Könyvkiadó), 1977.

K. Fazekas: "Bér-teljesitményalku a belsö munkaeröpiacon" (Bargaining about wages and performance in the inner labour market of the enterprises), in P. Galasi (ed): A munkaeröpiac szerkezete és müködése Magyarországon (Budapest, Közgazdasági és Jogi K., 1982), pp. 235-276.

J. Forslin, A. Sarapata, A.M. Whitehill, F. Adler and S. Mills (eds): Automation and Industrial Workers: A Fifteen Nation Study (Oxford, Pergamon Press, 1979 and 1981), 2 vols.

G. Friedmann: Le travail en miettes (Paris, Gallimard, 1956).

H. Hangnach and A. Schäfer: "German Democratic Republic", in ILO: New Forms of Work Organisation (Geneva, ILO, 1979), Vol. 2, pp. 3-23.

T.S. Hatchaturov: A szovjet gazdaság a kommunizmus építésének mai szakaszaban (The Soviet economy in the present phase of building communism) (Budapest, Közgazdasági és Jogi K., 1977). Russian original: Moscow, 1975.

L. Héthy: "Trade unions, shop stewards and participation in Hungary", in International Labour Review (Geneva, ILO), July-Aug. 1981, pp. 491-503.

L. Héthy: "The secondary economy in Hungary: its impact on industrial work and government efforts to control it", in Labour and Society (Geneva, ILO), July-Sept. 1982, pp. 243-253.

L. Héthy (ed): La situation des ouvriers dans l'usine: conditions de salaires et de travail de l'economie hongroise au cours des années 1970-1980 (Paris, ANACT, forthcoming), 210 p.

L. Héthy and C. Makó: A technika, a munkaszervezet és az ipari munka (Technology, work organisation and industrial work) (Budapest, Közgazdasági és Jogi Könyvkiadó, 1981), 303 p. [For a summary of its conceptual framework, see T. Huszár, K. Kulcsár and S. Szalai (eds): Hungarian Society and Marxist Sociology in the 1970s (Budapest, Corvina Press, 1978), pp. 120-133.]

L. Héthy and C. Makó: "Munkásmagatartások és a gazdasági szervezet" (Workers' behaviour and the economic organisation), in Akadémiai Kiadó (Budapest), 1972.

L. Héthy and C. Makó: "The workers and technological progress: the Hungarian Report", in J. Forslin et al. (eds): Automation and Industrial Workers: A Fifteen Nation Study (Oxford, Pergamon Press, 1981), Vol. 1, Part 2, pp. 233-255.

I. Kalinina and V. Roik: "Sokrashenie ruchnovo truda - zadacha technicneskaja, ekonomicheskaja, sochialnaja" (To decrease manual work - a technological economic and social task), in Sochialisticheskij Trud, 1983, No. 2.

V.V. Krevnevich: Vlijanie nauchno-techiceskovo progress na izmenenie strukturi rabochievo classa SSSR (The impact by scientific-technical progress on changes in the structure of working class in the USSR) (Moscow, Nauka, 1971), 389 p.

V.V. Krevnevich: "Avtomatizacia v sochialisticheskom obshestve kak uslovie povushenija sodershatelnosti truda i udovletvorennosti trudom" (Automation in the socialist society as a condition of job enrichment and job satisfaction), in Nauchnotechnicheskaja revolucija i rabochij klass v uslovijah socialisma (Prague, Institut Filosofii i Sociologii CsSzAN, 1978).

Kosanov et al.: "Upravlenie sochialnim razvitiem proizvodstvennih kollektivov" (The direction of the social development of collectives in production) (Nuka Kaz. SSR, Alma-Ata, 1975).

I. Lobanov and S.Z. Kuriteva: "Vlijánie mechanizacii i avtomatizacii na uslovija i sodierzanije truda" (The influence by mechanisation and automation on the conditions and content of work), in Sochialisticheskij Trud, 1982, No. 11.

C. Makó: "A munkaszervezet korszerüsitése az iparban" (Modernisation of work organisation in industry), in Társadalmi Szemle (Budapest, Magyar Szocialista Munkaspart) 1981.

C. Makó: "A müszaki-technikai fejlödés és a munkásbeállitottsagok" (Technological progress and the attitudes of workers), in Munkavédelmi Tanulmányok B. Sorozat 5 (Budapest, Táncsics Könyvkiadó, 1981).

R.K. Merton: "The machine, the worker and the engineer", in Industrial Sociology (New York, Harper and Row), 1964.

P.F. Petrochenko: Vlijánie nauchno-techniceskovo progressa na sodierzanie i organizaciju truda (Misl. Moskava, 1975).

Richta, Radovan et al.: Válaszutor a civilizáció (Civilisation at the junction) (Budapest, Kossuth Könyvkiadó, 1968).

A. Sarapata: "Polish automobile workers and automation", in J. Forslin et al. (eds): Automation and Industrial Workers: A Fifteen Nation Study (Oxford, Pergamon Press, 1979), Vol. 1, Part 1, pp. 118-153.

L. Sharikov: "Novaja technika-novie trebovanija i uslovija truda" (New technology - new requirements and conditions of work), in Sochialisticheskij Trud, 1982, No. 10.

A. Simonyi: "A központból a perifériára: strukturaváltás és munkásmagatartások egy üzemben" (From the centre to the periphery: structural changes and workers' behaviour in a factory), in Valóság (Budapest), 1978, No. 1, pp. 89-98.

J. Szczepanski: "A tudományos-technikai forradalom és a munkásosztály" (Scientific-technical revolution and the working class), in J. Szczepanski: A Szociológus Szemével (From the viewpoint of the sociologist) (Budapest, Gondolat, 1977), pp. 77-93. [Selected essays].

Szteriosz, Babanászisz, Denke and Géza: A tudományos-technikai forradalom és a munkások (Scientific-technical revolution and the workers) (Budapest, Közgazdasági és Jogi Könyvkiadó, 1979), 390 p.

J. Timár (ed): Munkagazdaságtan (Labour economics) (Budapest, Közgazdasági és Jogi Könyvkiadó, 1981).

A. Touraine: L'organisation du travail aux usines Renault (Paris, CNRS, 1955).

A. Touraine: L'organisation du travail et l'entreprise: Traite de sociologie du travail (Colin, Armand, 3rd edition, 1970).

Ussenin, Krevnevich, Maslov, Notchevnik, Podkosov, Kolbanovsky, Denisovsky and Sarovsky: "Soviet workers and automation of the production process", in J. Forslin et al. (eds): Automation and Industrial Workers: A Fifteen Nation Study (Oxford, Pergamon Press, 1979), Vol. 1, Part 1, pp. 154-198.

C.R. Walker and H.G. Guest: The Man on the Assembly Line (Cambridge, Harvard University Press, 1952).

J. Woodward: Industrial Organizations, Theory and Practice (London, Oxford University Press, 1965).

A.G. Adravomyslov, V.N. Rozhin and V.A. Jadov: Man and His Work (New York, International Arts and Sciences Press, 1970).

N. Zemlanskij, N. Kuznecov and A. Svataev: "Novie formie organizacii truda na avtomaticéskih liniah", in Sochialisticheski Trud, 1977, No. 2.

V.P. Zinchenko and V.M. Munipov: "A termelés automatizálásának ergonomiai szempontjai" (Ergonomic aspects of the automation of production), in Ergónomia (Budapest), 1978, Vol. 11, No. 1, pp. 1-6.

WORK ORGANISATION AND OCCUPATIONAL STRESS

C.L. Cooper[1]

I. Nature of stress

1. Stress (a word derived from Latin) was used popularly in the seventeenth century to mean hardship, straits, adversity, or affliction. During the late eighteenth century its use evolved to denote force, pressure, strain or strong effort, with reference primarily to a person or to a person's organs or mental powers (Hinkle, 1973).

2. The idea that stress contributes to long-term ill health (rather than merely short-term discomfort implicit in the above definition) can also be found early on in the concept's development.

3. Hans Selye (1946) was one of the first to try to explain the process of stress-related illness with his "general adaptation syndrome" theory. In it he described three stages an individual encounters in stressful situations:

(i) The alarm reaction in which an initial shock phase of lowered resistance is followed by countershock during which the individual's defence mechanisms become active;

(ii) resistance, the stage of maximum adaptation and, hopefully, successful return to equilibrium for the individual. If, however, the stressor continues or the defence does not work, he will move on to the third stage;

(iii) exhaustion, when adaptive mechanisms collapse.

II. The costs of occupational stress

4. Stress-related illnesses such as coronary heart disease have been on a steady upward trend over the past couple of decades in countries where few planned interventions have been implemented in dealing with the pressures

of work life. In England and Wales, for example, the death rate in men between 35 and 44 nearly doubled between 1950 and 1972, and has increased much more rapidly than that of older age ranges (e.g. 45-54). By 1973, 41 per cent of all deaths in the age group 35-44 were due to cardiovascular disease, with nearly 30 per cent due to cardiac heart disease. In fact, in 1976 the American Heart Association estimated the cost of cardiovascular disease in the United States at $26.7 billion a year. In addition to the more extreme forms of stress-related illnesses, there has been an increase in other possible stress manifestations such as alcoholism, where admissions to alcoholism units in United Kingdom hospitals, for example, has increased from roughly under 6,000 in 1966 to over 8,000 in 1974, and industrial accidents and short-term illnesses (through certified and uncertified sick leaves), with an estimated 300 million working days lost at a cost of £55 million in United Kingdom national insurance and supplementary benefits payments alone. The total cost to industry of all forms of stress-related illness and other manifestations, a large slice of which can be attributed directly or indirectly to the working environment, must be enormous, beyond the scope of most cost accountants to begin to calculate. Some Americans estimate that it may represent in the order of 1-3 per cent of GNP in the United States (McMurrya, 1973).

5. The commonly held belief about coronary heart disease, peptic ulcers and other stress-related illnesses in terms of occupations and work is that they are found predominantly among "professions", i.e. that they are the "boss' disease". This myth can be firmly put to rest: work pressure affects all workers and the sources and manifestations of stress vary from occupation to occupation.

Table 1: Deaths by major causes and types of occupations, 1970-72 (standardised mortality rates = 100)

Causes of death in persons aged 15-64 (males)	Professional and similar	Intermediate	Skilled non-manual	Skilled manual	Partly skilled	Unskilled
Trachea, bronchus and lung cancer	53	68	84	118	123	143
Prostate cancer	91	89	99	115	106	115
Ischaemic heart disease	88	91	114	107	108	111
Other forms of heart disease	69	75	94	100	121	157
Cerebrovascular disease	80	86	98	106	111	136
Pneumonia	41	53	78	92	115	195
Bronchitis, emphysema and asthma	36	51	82	113	128	188
Accidents other than motor vehicle	58	64	53	97	128	225
All causes	77	81	99	106	114	137

Source: United Kingdom Office of Population Censuses and Surveys. Registrar General's annual estimates of the population of England and Wales and of local authority areas, 1970.

6. If we examine table 1, we can see that frequencies of deaths due to major causes in the United Kingdom working population increase as we move from professional and white-collar jobs down to the unskilled. This applies to both stress-related illnesses such as ischaemic heart disease and to other illnesses such as pneumonia and prostate cancer. These statistics are very similar to mortality data from the United States and other developed countries. In terms of almost all the major and many of the minor causes of death among persons in the working population age groups, the blue-collar and unskilled are at greater risk than the white-collar and professional groups. This extends not only to mortality statistics but also to morbidity data as well. It can be seen in table 2 that many blue-collar jobs show a greater number of restricted activity days and consultations with general practitioners than do white-collar workers in the United Kingdom. But does this mean that professional and managerial workers are not stress prone, that their occupations and life style minimise their vulnerability to stressors at work (and at home) and, consequently, to minor and serious illness?

7. Cherry (1978), in a large-scale study on stress, anxiety and work among 1,415 men, found that a higher proportion of professional and white-collar workers reported nervous strain at work than did skilled, semi-skilled and unskilled manual workers. Looking at her sample in terms of the United Kingdom Office of Population Censuses and Surveys categories, she found the following percentage of workers reporting nervous debility and strain at work: professional 53.8 per cent, intermediate non-manual 56.9 per cent, skilled non-manual 15.3 per cent and unskilled manual 10.3 per cent. This may indicate only that white-collar and professional groups differ from blue-collar occupations in their reactions to stress, that is, that the former reflect the pressures of work in mental illness, whereas the latter do so in physical symptoms and illness.

Table 2: Acute sickness and consultations with general medical practitioners, 1974-1975

	Average number of restricted activity days per person per year (males)			Average number of consultations per person per year (males)		
	15-44	45-64	All ages	15-44	45-64	All ages
Professional	9	16	12	2.1	2.7	2.7
Employers and managers	11	13	14	1.8	2.4	2.7
Intermediate and junior non-manual	10	21	15	2.0	4.3	3.1
Skilled manual and own account non-professional	15	24	17	2.8	4.0	3.2
Semi-skilled manual and personal service	16	23	18	2.7	4.5	3.7
Unskilled manual	21	28	20	3.5	4.8	3.6
All persons	13	21	16	2.4	3.8	3.1

Source: General Household Survey, United Kingdom, HMSO, 1974 and 1975.

III. Relationship between work organisation and occupational stress

8. There is considerable evidence from studies in the workplace (Margolis et al., 1974) to suggest that occupational stressors are contributory factors in coronary heart disease (CHD) and other stress-related illnesses. Occupational stress is here defined as negative environmental factors or stressors (e.g. work overload, role conflict/ambiguity, poor working conditions) associated with a particular job. Inherent in the concept of occupational stress is the interaction of the person with his environment, giving rise to coping or maladaptive behaviour and ultimately to stress-related disease (Cooper, 1983). Lofquist and Dawis (1969) have labelled this interaction "the person-environment fit". Therefore, it is important to identify those factors within the occupational environment which impinge on the individual, creating conditions for stress-related illness and potential CHD. (See figure 1)

9. Most research indicates that, depending on the particular job and organisation, one or some combination of the sources of stress in the above model together with certain personality traits, may be predictive of a variety of stress manifestations, such as coronary heart disease, mental ill health, job dissatisfaction, marital disharmony, exessive alcoholic intake or other drug taking, etc.

10. The six major sources of occupational stress will be discussed: factors intrinsic to the job; role in the organisation; career development; relationship at work; organisational structure and climate; and home work interface.

A. Factors intrinsic to the job

11. Sources of stress intrinsic to the job across a variety of occupations include: (1) poor physical working conditions, (2) shift work, (3) work overload, (4) work underload, (5) physical danger, (6) person-environment fit (P-E) and job satisfaction (Cooper and Marshall, 1976).

12. Poor physical working conditions can enhance stress at work. In regard to nuclear power plant operators, for example, Otway and Misenta (1980) believe that the design of the control room itself is an important variable in terms of worker stress. They propose that control room designs need to be updated, requiring more sophisticated

ergonomic designs. Furthermore, they refer to the Kemeny et et al. study which highlighted an important stress factor in the Three Mile Island Accident as being the distraction caused by excessive emergency alarms. In a study carried out by Kelly and Cooper (1981) on the stressors associated with casting in a steel manufacturing plant, they found poor physical working conditions to be a major stressor. (See table 3) Many of the stressors were concentrated in the physical aspects of noise, fumes and to a lesser extent, heat plus the social and psychological consequences of isolation and interpersonal tension. A further possible source of stress was seen to be in the lack of job satisfaction, particularly arising from the stressors above, and partially endemic to the nature of casting of liquid steel, in a continuous process lasting some 70 minutes. For 75 per cent of this time-cycle the casters were exposed to, and by the nature of their task, unable to move away from, very high levels of noise (up to 110 dB for much of the time) and periodic and unpleasant air pollution caused by the activities of other workers and machines in their proximity. These conditions necessitate the wearing of ear protection, in the form of ear muffs or cotton wool swabs which effectively isolate the wearer from those around him.

13. Marcson (1970) and Shepart (1971) found, for example, that physical health is adversely affected by repetitive and dehumanising environments (e.g. paced assembly lines). In addition, Kritsikis, Heinemann and Eitner (1968), in a study of 150 men with angina pectoris in a population of over 4,000 industrial workers in Berlin, reported that more of these workers came from work environments employing conveyor-line systems than in any other work technology.

14. Shift work. Numerous occupational studies have found that shift work is a common occupational stressor as well as affecting neurophysiological rhythms, such as blood temperature, metabolic rate, blood sugar levels, mental efficiency and work motivation, which may ultimately result in stress-related disease (Selye, 1976). A particular occupational study by Cobb and Rose (1973) on air traffic controllers (a particularly strenuous occupation) found four times the prevalence of hypertension and also more mild diabetes and peptic ulcers among the subjects than in their control group of second class airmen. Although these authors identified other job stressors as being instrumental in the causation of these stress-related maladies, a major job stressor was isolated as shift work. Nevertheless, although there are stressors associated with shift work, one needs to take note of Selye's conclusion on the issue. He suggests

Figure 1

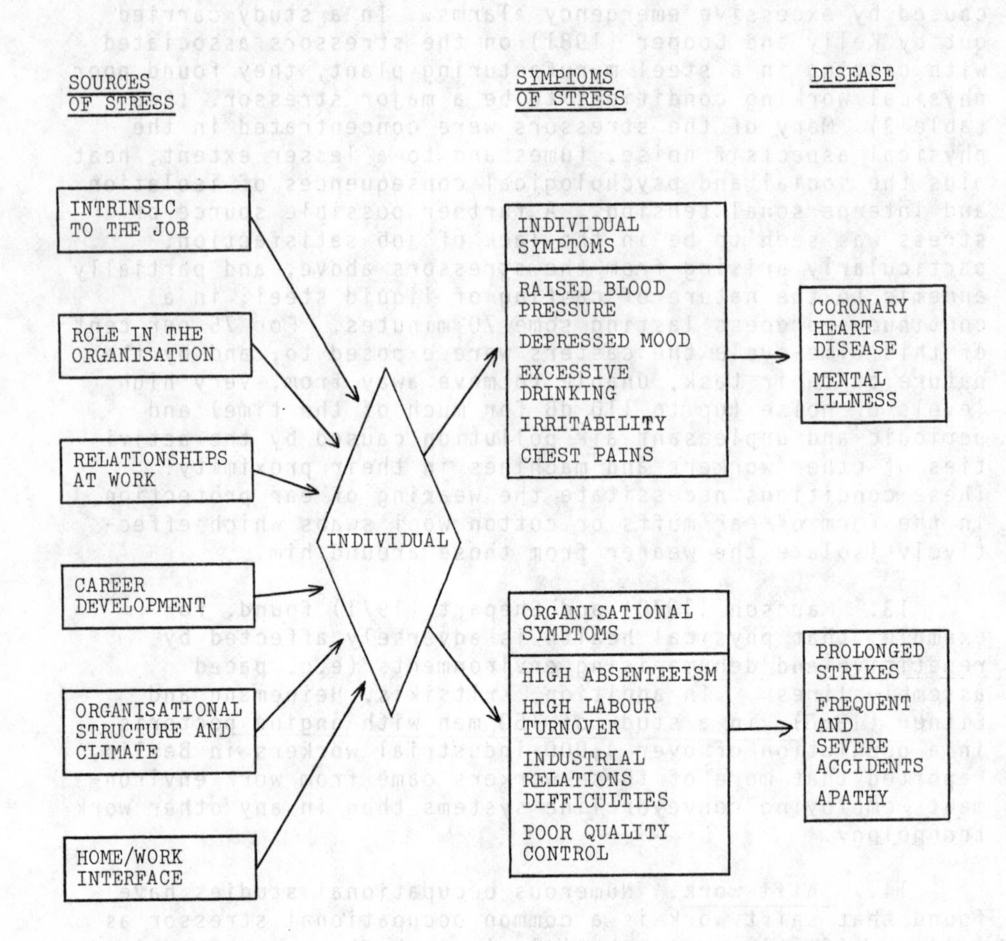

SOURCES OF STRESS
SYMPTOMS OF STRESS
DISEASE
INTRINSIC TO THE JOB
ROLE IN THE ORGANISATION
RELATIONSHIPS AT WORK
CAREER DEVELOPMENT
ORGANISATIONAL STRUCTURE AND CLIMATE
HOME/WORK INTERFACE
INDIVIDUAL
INDIVIDUAL SYMPTOMS
RAISED BLOOD PRESSURE
DEPRESSED MOOD
EXCESSIVE DRINKING
IRRITABILITY
CHEST PAINS
ORGANISATIONAL SYMPTOMS
HIGH ABSENTEEISM
HIGH LABOUR TURNOVER
INDUSTRIAL RELATIONS DIFFICULTIES
POOR QUALITY CONTROL
CORONARY HEART DISEASE
MENTAL ILLNESS
PROLONGED STRIKES
FREQUENT AND SEVERE ACCIDENTS
APATHY

Table 3: Stressors in the job of steel casting perceived by the casting crew

Stressor item	% of respondents
1. Physical environment factors	
Noise	100
Dirt/dust	88.2
Vibration	29.4
Poor accommodation for rest periods	29.4
Lack of toilet facilities nearby	23.5
Heat	5.9
2. Relations with other groups	
Pressure of work from furnace workers	35.3
Thoughtless actions by other workers	35.3
Poor view of casters by other groups	17.7
3. Psychological/personal factors	
Immobility on the job	41.2
Isolated when casting	23.5
Conditions cause lost tempers, worn nerves	23.5
4. Health and safety factors	
Danger hazards	17.7
Deterioration in physical health	5.9
Need to wear protective clothing, headphones	5.9

Source: Kelly and Cooper, 1981.

that most investigations agree that shift work becomes physically less stressful as individuals can (and often do) habituate to the condition. Even so, being "excluded from society" is a common complaint among shift workers.

15. Job overload. French and Caplan (1970, 1972) see work overload as being either quantitative (i.e. having too much to do) or qualitative (i.e. being too difficult) and certain behavioural malfunctions have been associated with job overload. For example, the French and Caplan study indicated a relationship with quantitative overload and cigarette smoking (an important symptom in relation to coronary heart disease) and Margolis et al., in their sample of 1,500 employees, found that job overload was associated with such stress-related symptoms and lowered self-esteem, low work motivation and escapist drinking.

16. In a study of 100 young coronary patients, Russek and Zohman (1958) found that 25 per cent had been working at two jobs and an additional 45 per cent had worked at jobs which required (due to work overload) 60 or more hours per week. They add that although prolonged emotional strain preceded the attack in 91 per cent of the cases, similar stress was only observed in 20 per cent of the controls. Breslow and Buell have also reported findings which support a relationship between hours of work and death from coronary heart disease. In an investigation of mortality rates of men in California, they observed that workers in light industry under the age of 45, who are on the job more than 48 hours a week, have twice the risk of death from CHD compared with similar workers working 40 or under hours a week. There is also some evidence that (for some occupations) qualitative overload is a source of stress. French, Tupper and Mueller (1965) looked at qualitative and quantitative work overload in a large university. They used questionnaires, interviews and medical examinations to obtain data on risk factors associated with CHD for 122 university administrators and professors. They found that one symptom of stress, low self-esteem, was related to work overload but that this was different for the two occupational groupings. Qualitative overload was not significantly linked to low self-esteem among the administrators but was significantly correlated for the professors. The greater the "quality" of work expected of the professor, the lower the self-esteem. They also found that qualitative and quantitative overload were correlated to achievement orientation. And more interestingly, in a follow-up study, achievement orientation correlated very strongly with serum uric acid (Brooks and Mueller, 1966). Several other studies have reported an association of qualitative work overload with cholesterol level: a tax deadline for accounts (Friedman,

Rosenman and Carroll, 1958), medical students performing a medical examination under observation (Dreyfuss and Czackes, 1959), etc. French and Caplan summarise this research by suggesting that both qualitative and quantitative overload produce at least nine different symptoms of psychological and physical strain: job dissatisfaction, job tension, lower self-esteem, threat, embarrassment, high cholesterol levels, increased heart rate, skin resistance, and more smoking.

17. Job underload. Job underload associated with repetitive, routine, boring and understimulating work environments (e.g. paced assembly lines) has been associated with ill health (Cox, 1980). Moreover, in certain jobs, such as policing and operating nuclear power plants, periods of boredom have to be accepted, along with the possibility that one's duties may suddenly be disrupted due to an emergency situation (Davidson and Veno, 1980). This can give a sudden jolt to the physical and mental state of the employee and have a subsequent detrimental effect on health (McCrae, Costa and Bosse, 1978). Furthermore, boredom and disinterest in the job may reduce the employee's response to emergencies (Davidson and Veno, 1980).

18. Physical danger. There are certain occupations which have been isolated as being high risk in terms of danger, e.g. police, mine workers, soldiers and firemen (Kasl, 1973). However, stress induced by the uncertainty of physical danger events is often substantially relieved if the employee feels adequately trained and equipped to cope with emergency situations.

19. One high risk occupation is that of the bomb disposal officer. Cooper carried out an investigation on 40 bomb disposal officers in Northern Ireland, exploring the psychometric differences between successful and unsuccessful ones using the Dynamic Personality Inventory (DPI) (See table 4). It was found that there were nine significant differences between the two groups, which indicated a consistent social behavioural pattern. The successful bomb disposal expert seems to have a low level affiliation and affectation motivation (factors H, Od, Tl), difficulty in forming and maintaining close personal relations (factors C, Pn, Od, Tl) and a tendency towards non-conformity, relying less on conventional values and judgements (factors Ac, H, Od). Successful bomb disposal operators, therefore seem to be social isolates preferring to work on their own and with "things" as opposed to people. The lack of concern with social relationships and the ability to maintain "personal distance" may ease the inevitable tensions that arise as bomb disposal colleagues are killed, maimed and injured in the course of their work.

20. P-E fit and job satisfaction. Finally, a measure of job satisfaction and related variables which deserves mention is the measure known as person-environment fit (P-E) (Caplan et al., 1975). According to McMichael, P-E fit can be defined as an interaction between an individual's psychosocial characteristics and objective environmental work conditions. Consequently, a score of P-E fit can be attained by subtracting the amount/degree of a particular job factor (e.g. work load) preferred by a person from the actual amount in that same person's job environment. The overall hypothesis is that stress can occur and result in such problems as anxiety, depression, job dissatisfaction and physiological maladies if there is a P-E misfit.

B. Role in the organisation

21. Another potential source of occupational stress is associated with a person's role at work. In particular, he/she may experience role conflict, which exists when an individual in a work role is torn by conflicting job demands, or doing things he/she really does not want to do or does not think is part of the job specification. The most frequent manifestation of this is when a person is caught between two groups of people who demand different kinds of behaviour or expect that the job should entail different kinds of behaviour or functions. Shirom et al. collected data on 762 male Kibbutz members aged 30 and above, drawn from 13 Kibbutzim throughout Israel. They examined the relationships between CHD (myocardial infarcation, angina pectoris, and coronary insufficiency), abnormal electro-cardiographic readings, CHD risk factors (systolic blood pressure, pulse rate, serum cholesterol levels, etc.) and potential sources of occupational stress (work overload, role ambiguity, role conflict, lack of physical activity). Their data was broken down by occupational groups: agricultural workers, factory groups, craftsmen, and white-collar workers. It was found that there was a significant relationship between role conflict and CHD (specifically, abnormal electro-cardiographic readings), but for the white-collar workers only. In fact, as we moved down the ladder, from occupations requiring great physical exertions (e.g. agriculture) to least (e.g. white-collar), the greater was the relationship between role ambiguity/conflict and abnormal electro-cardiographic findings. Role conflict was also significantly related to an index of ponderosity (excessive weight for age and height). It was also found that as we go from occupations involving excessive physical activities to those with less activity, CHD increased significantly. The inference drawn from this is that as we move more towards clerical, managerial and

Table 4: Means, standard deviations (SD) and t values for significance of differences between successful and control bomb disposal experts on DPI scales

DPI scales	M_S	SD	M_C	SD	t
H	6.25	1.37	7.60	2.34	2.21*
W_p	5.65	2.66	6.90	1.91	1.70
W_s	5.15	2.96	5.70	2.67	0.61
O	4.70	2.03	6.90	1.61	3.79ǂ
O_A	6.70	1.83	6.30	1.92	0.67
O_d	5.20	1.85	6.75	1.20	3.13**
O_m	5.20	1.82	5.55	1.84	0.60
O_v	4.95	1.98	4.60	1.87	0.57
O_j	5.95	2.37	5.70	1.94	0.36
O_u	5.30	1.92	5.25	1.68	0.08
A_n	5.25	1.80	5.50	1.79	0.43
A_d	5.85	2.36	7.15	1.95	1.89
A_c	5.20	1.70	6.40	1.81	2.15*
A_a	5.70	1.89	6.65	1.75	1.64
A_s	6.25	2.07	6.55	1.76	0.49
A_i	5.55	1.73	4.80	2.23	1.18
P	5.95	2.11	6.50	1.79	0.88
P_n	5.40	2.13	7.40	1.56	3.37**
P_e	6.25	1.88	7.10	1.83	1.44
P_a	5.70	2.00	6.10	1.99	0.63
P_h	5.35	2.03	5.55	2.06	0.30
P_l	6.20	2.23	5.65	2.18	0.78
P_i	6.95	2.01	5.80	1.96	1.82
S	5.95	1.46	6.05	1.35	0.22
TI	5.55	1.53	7.25	1.99	3.01**
CI	5.55	1.81	6.75	2.07	1.94
M	7.25	2.26	7.35	2.39	0.13
F	4.75	1.94	7.40	1.46	4.86ǂ
MF	6.20	1.88	7.70	1.92	2.49*
SA	6.10	1.94	6.80	2.11	1.08
C	5.70	2.03	7.35	2.00	2.58*
EP	6.55	1.73	6.20	1.67	0.65
EI	7.10	1.83	7.30	2.03	0.32

* P<.05
ǂ P<.01
** P<.001

Source: Cooper, C.L., Journal of Occupational Medicine, 24(9), 1982

professional occupations, we may be increasing the likelihood of occupational stress due to identity and other interpersonal dynamics and less to the physical conditions of work.

22. There are a number of studies which relate occupational level to CHD and mental ill-health (MIH), of which Marks (1967) provides an excellent review. The majority of these studies support the proposition that risk of CHD rises with occupational level (Syme, Hyman and Enterline, 1964). Substantial national analyses of both British and American mortality data lend support to these studies. Not all researchers, however, are in agreement. Pell and D'Alonzo, in a highly consistent longitudinal study of Dupont employees, found that incidence of myocardinal infarcation was inversely related to salary roll level. Stamler, Kjelsberg and Hall and Bainton and Peterson also came up with contradictory results. A further group of researchers have added confusion by finding no relationship between CHD and occupation (Berkson; Spain; Paul).

23. Another important potential stressor associated with one's organisational role is "responsibility for people". One can differentiate here between "responsibility for people" and "responsibility for things" (equipment, budgets, etc.). Wardwell et al. found that responsibility for people was significantly more likely to lead to CHD than responsibility for things. Increased responsibility for people frequently means that one has to spend more time interacting with others, attending meetings, working alone and, in consequence, as in the Goddard study, more time in trying to meet deadline pressures and schedules. Pincherle also found this in a United Kingdom study of 2,000 executives attending a medical centre for a medical check-up. Of the 1,200 managers sent by their companies for their annual examination, there was evidence of physical stress being linked to age and level of responsibility; the older and more responsible the executive, the greater the probability of the presence of CHD risk factors or symptoms.

24. Other research (Tethune, 1963) has also established this link. The relationship between age and stress-related illness could be explained, however, by the fact that as the executive gets older he may be troubled by stressors other than increased responsibility, for example, as Eaton suggests by (1) a recognition that further advancement is unlikely; (2) increasing isolation and narrowing of interests and (3) an awareness of approaching retirement. Nevertheless, the finding by French and Caplan in the Goddard study does indicate that responsibility for people must play some part in the process of stress, particularly

for clerical, managerial and professional workers. They found that responsibility for people was significantly related to heavy smoking, diastolic blood pressure and serum cholesterol levels, the more the individual had responsibility for things as opposed to people, the lower were each of these CHD factors.

C. Career development

25. The next group of environmental stressors is related to career development which refers to "the impact of over-promotion, under-promotion, status incongruence, lack of job security, thwarted ambition ...". Status congruency or the degree to which there is job advancement (including pay grade advancement) was found by Erickson, Pugh and Gunderson in their large sample of Navy employees to be positively related to military effectiveness and negatively related to the incidence of psychiatric disorders. However, in terms of pay, Otway and Misenta postulate that large increases in workers pay would not necessarily mean simultaneous increases in job satisfaction and might result in personnel remaining in jobs which no longer give them satisfaction.

26. Career development blockages are most noticeable among women managers as a study by Cooper and Davidson revealed recently. In this investigation the authors collected data from over 700 female managers and 250 male managers at all levels of the organisational heirarchy and from among several hundred companies. It was found that women suffered significantly more than men on a range of organisational stressors, but the most damaging to their health and job satisfaction were the ones associated with career development and allied stressors (e.g. sex discrimination in promotion, inadequate training, male colleagues trained more favourably, not enough delegation to women).

D. Relationships at work

27. Relationships at work, which include the nature of relationships and social support from one's colleagues, boss and subordinates, have also been related to job stress (Payne, 1980). According to French and Caplan, poor relationships with other members of an organisation may be precipitated by role ambiguity in the organisation, which in turn may produce psychological strain in the form of low job satisfaction. Moreover, Caplan et al. (1975) found that strong social support from peers relieved job strain and also served to condition the effects of job stress on

cortisone, blood pressure, glucose and the number of cigarettes smoked as well as cessation of cigarette smoking. It is interesting to note that in air traffic controllers, greater help and social support were provided by friends and colleagues than by those in supervisory positions (French and Caplan, 1970).

28. In addition, where male executives had problems, they were associated with problems in relationships as Cooper and Melhuish discovered in their study of 196 very senior male executives. It was found that male executives' personality pre-dispositions (e.g. outgoing, tough minded, etc.) and their relationships at work were central to their increased risk of high blood pressure. They were particularly vulnerable to the stresses of poor relationships with subordinates and colleagues, lack of personal support at home and work, and to the conflicts between their own values and those of the organisation.

Table 5: Stepwise multiple regression analysis of work and personality stressors and raised blood pressures (diastolic and systolic) among male executives

Step	Stressor variable	Multiple R	R^2	R^2 change
1	Outgoing personality (Factor A)	0.227	0.051	0.051
2	Tough-minded (Factor 1)	0.319	0.102	0.051
3	Conflict between personal values and company	0.377	0.142	0.040
4	Poor relationships with subordinates and colleagues	0.422	0.178	0.036
5	Little social support from home and work	0.487	0.237	0.059
6	Self-sufficient personality (Factor C2)	0.518	0.269	0.032
7	Forthright personality (Factor N)	0.556	0.309	0.040
8	Venturesome personality (Factor H)	0.595	0.354	0.045
9	Assertive personality (Factor E)	0.618	0.382	0.028

Source: C.L. Cooper and A. Melhuish : Journal of occupational medicine, 22(10), 1980.

E. Organisational structure and climate

29. Another potential source of occupational stress is related to organisational structure and climate, which includes such factors as office politics, lack of effective consultation, lack of participation in the decision-making process and restrictions on behaviour. Margolis et al. and French and Caplan found that greater participation led to higher productivity, improved performance, lower staff turnover and lower levels of physical and mental illness (including such stress-related behaviours as escapist drinking and heavy smoking).

F. Home: work pressures

30. Another danger of the current economic situation is the effect that work pressures (such as fear of job loss, blocked ambition, work overload and so on) have on the families of the employees. At the very best of times young managers, for example, face the inevitable conflict between organisational and family demands during the early build-up of their careers, as the British Institute of Management survey (Beattie et al. 1974), entitled "The Management Threshold" illustrates. But during a crisis of the sort we are currently experiencing the problems increase in geometrical proportion as managers strive to cope with some of their basic economic and security needs. As Pahl and Pahl suggest in their book on Managers and their wives, most male managers, under normal circumstances, find a home a refuge from the competitive and demanding environment of work and a place where they can get support and comfort. However, when there is a career crisis (or stress from job insecurity as many managers are now facing), the tensions male managers bring with them into the family affect their wives and home environment in a way that may not meet their "sanctuary" expectations. It may be very difficult for example, for a wife to provide a kind of supportive domestic scene her husband requires at a time when she is beginning to feel insecure, when she is worried about the family's economic, educational and social future.

31. Dual career stress. It is difficult enough for a housebound wife to support her breadwinning husband and at the same time cope with family demands, but increasingly women are following full-time careers as well. According to the United States Department of Labor the "typical American family" with a working husband, a homemaker wife and two children now makes up only 7 per cent of the nation's families. In fact, in 1975, 45 per cent of all married women were working as were 37 per cent of women with children

under six; in 1960 the comparable figures were 31 per cent and 19 per cent respectively. It is claimed by many psychologists and sociologists that the development of dual-career families is a significant factor in the very large increase in the divorce rate over the last 10 years in the United States and countries in Western Europe (Payne,1980).

32. The problems this creates for both workers are enormous, it affects almost all aspects of their lives at work. For example, managers are expected, as part of their job, to be mobile, that is, to be readily available for job transfers, both within and between countries. Indeed a manager's promotional prospects depend wholly on his/her availability and willingness to accept promotional moves. In the 1980s and 1990s, as women themselves pursue full-time careers with increasing responsibilities as opposed to "support" type jobs the prospect of managers being available for rapid deployment will decrease substantially. In the past, male managers have, with few exceptions, accepted promotional moves almost without family discussion. Future such decisions will create major obstacles for both bread-winners in the family. We are already seeing this happen throughout Europe and the United States, and this is particularly exacerbated by the fact that corporations have not adapted to this changing social phenomenon. Few facilities, as yet, are available in companies to help either of the dual-career members of the family unit.

IV. Coping with stress

A. Individual differences in coping with stress

33. Sources of pressure at work evoke different reactions from different people. Some people are better able to cope with these stressors than others, they adapt their behaviour in a way that meets the environmental challenge. On the other hand some people are more characterologically predisposed to stress, that is, they are unable to cope or adapt to the stress-provoking situation. A great deal of research has been done on the individual differences associated with stress-related diseases, particularly CHD. There have been two principal directions of research in this area: one has concentrated on examining the relationship between various psychometric measures (primarily using Minnesota Multiphasic Personality Inventory and Sixteen Personality Factor Questionnaire) and stress-related disease (primarily CHD); and the other stress or coronary-prone

behaviour patterns and the incidence of disease. Jenkins (1971) provides an extensive review of these studies. In the first category, there were six studies which utilised the MMPI. The result of these six studies seems to be that before their illness, patients with coronary disease differ from persons who remain healthy on several MMPI scales, particularly those in the "neurotic" triad of hypochondriasis (Hs), depression (D), and hysteria (Hy). The occurence of manifest CHD increases the deviation of patient's MMPI scores further and, in addition, there is ego defence breakdown. As Jenkins summarises "patients with fatal disease tend to show greater neuroticism (particularly depression) in prospective MMPI's than those who incur and survive coronary disease". There are three major studies utilising the 16PF. All three of these report emotional instability (low scale C), particularly for patients with angina pectoris. Two studies report high conformity and submissiveness (Factor E) and desurgency/seriousness (Factor F), and two report high self-sufficiency (Factor Q2). Bakker's angina patients are similar to Finn's sample with CHD, in manifesting shyness (Factor H) and apprehensiveness (Factor O). The results from all three studies portray the patients with CHD or related illness as emotionally unstable and introverted, which is consistent with the six MMPI studies. The limitation of these studies is that they are, on balance, restrospective. That is, that anxiety and neuroticism may well be reactions to CHD and other stress-related illnesses rather than precursors of it. Paffenbarger, Wolf and Notkin (1966) did an interesting prospective study, in which they linked university psychometric data on students with death certificates filed years later. They found a number of significant precursors to fatal CHD, one of which was a high anxiety/neuroticism score for the fatal cases.

34. The other research approach to individual differences began with the work of Friedman and Rosenman in the early sixties and developed later showing a relationship between behavioural patterns and the prevalence of CHD. They found that individuals manifesting certain behavioural traits were significantly more at risk to CHD. These individuals were later referred to as the "coronary-prone behaviour pattern Type A" as distinct from Type B (low risk to CHD). Type A was found to be the overt behavioural syndrome or style of living characterised by "extremes of competitiveness, striving for achievement, aggressiveness, haste, impatience, restlessness, hyper-alertness, explosiveness of speech, tenseness of facial musculature and feelings of being under pressure of time and under the challenge of responsibility". It was suggested that "people having this particular behavioural pattern were often so deeply involved and committed to their work that other aspects of their lives were relatively neglected" (Jenkins, 1971b). In the

early studies, persons were designated as Type A or Type B on the basis of clinical judgements of doctors and psychologists or peer ratings. These studies found higher incidence of CHD among Type A than Type B.

35. Many of the inherent methodological weaknesses of this approach were overcome by the classic Western Collaborative Group Study (Rosenman et al., 1964, 1966). It was a prospective (as opposed to the earlier retrospective studies) national sample of over 3,440 men free of CHD. All these men were rated Type A or B by psychiatrists after intensive interviews, without knowledge of any biological data about them and without the individuals being seen by a cardiologist. Diagnosis was made by an electrocardiographer and independent medical internist, who were not informed about the subjects' behavioural patterns. They found the following results: after 2 1/2 years of the start of the study, Type A men between the ages of 39-49 and 50-59 had 6.5 and 1.9 times respectively the incidence of CHD than Type B men. They also had the following risk factors of elevated serum cholesterol levels, elevated bitalipoproteins, decreased blood clotting time, and elevated daytime excretion of norepinephrine. After 4 1/2 years of the follow-up observation in the study, the _same_ relationship of behavioural pattern and incidence of CHD was found. In terms of the clinical manifestations of CHD, individuals exhibiting Type A behavioural patterns had significantly more incidence of acute myocardial infarcation (and of clinically unrecognised myocardial infarcation) and angina pectoris. Rosenman et al. (1967) also found that the risk of recurrent and fatal myocardial infarcation was significantly related to Type A characteristics.

36. Quinlan and his colleagues (1969) found the same results among Trappist and Benedictine monks. Monks judged to be Type A coronary-prone cases (by a double-blind procedure) had 2.3 times the prevalence of angina and 4.3 times the prevalence of infarcation as compared to monks judged Type B. Many other studies have been conducted with roughly the same findings.

37. French and Caplan (1970) conclude their review of this evidence by saying "such wealth of findings makes it hard to ignore Type A as a relevant syndrome". It can be seen, therefore, that psychometric and behavioural data on individual differences play a crucial role in the Person-Environment Fit paradigm and ultimately in the manifestation of stress-related disease.

B. Ways of coping with stress

38. The evidence in the foregoing discussion suggests that certain occupational stressors may be major contributors to coronary risk. However, the task of removing harmful stress from the workplace is not as easy as it first appears. In the majority of cases occupational stressors and the experience of stress are directly linked to job satisfaction; hence the paradox that a factor which is a major source of satisfaction can, at the same time, increase the individual's risk of CHD. Therefore, the task of coping strategies is to prevent the deleterious effects of stress, whilst not eliminating it as a source of satisfaction. Selye emphasises that "we must not suppress stress in all its forms, but diminish distress and facilitate eustress", which he defines as "the satisfactory feeling that comes from the accomplishment of tasks that we consider worthwhile".

39. There is evidence that different people are able to sustain stress with differing degrees of effectiveness; it has been found that people with certain types of behaviour patterns are more likely to be at coronary risk than others (Friedman and Rosenman, 1974). The "subjective" nature of stress means that levels of stress will differ between people according to their circumstances; similarly different coping techniques will be more or less effective according to the individual (Welford, 1974).

40. A number of possible methods have been studied under laboratory conditions and in the workplace, which may help us to understand and cope with stress. First, there are the "cognitive coping strategies" open to the individual for stress reduction. Second, there are the tension-reducing or relaxation techniques. And finally, there is the social support one can get from one's work and family group.

41. Cognitive coping with stress. Lazarus suggests that in a stressful situation the individual uses an active coping strategy which he believes will counter the threat. If active coping strategies fail, then cognitive coping is induced in the form of "situation redefinition" or "denial". This has been interpreted in another way by identifying response categories in terms of behavioural, cognitive or decisional control over the impending stressor (Averill, 1973). Behavioural control is the availability of a response which may directly influence or modify the objective characteristics of a threatening event, whereas cognitive control is the way in which an event is interpreted or appraised. Decisional control is the opportunity to choose among various courses of action. Sells believes that the

lack of control over a stressful stimulus is a necessary if not sufficient condition for stress. Therefore, if personal control of the objective conditions is lost, it can be regained in subjective terms by incorporating a potentially threatening event into a cognitive plan, thereby reducing anxiety (Mandlar et al., 1966).

42. A number of potential environmental stressors have been used in experiments to test the effectiveness of cognitive coping (Lazarus et al., 1965), and intellectualisation reduced stress, as measured by physiological reactions. This work was criticised for its methodological weaknesses by Holmes and Houston who tested the nature and effectiveness of more specific cognitive strategies for handling stress. They employed two strategies of redefinition and affective isolation, with the hypothesis that these would evidence less response to stress than subjects not instructed in the use of these strategies. The experimental results revealed that the cognitive coping subjects were more effective in reducing stress than the control subjects.

43. Houston (1973) using groups of high trait and low trait denial subjects found that in stressful situations high trait denial subjects had significantly less physiological arousal, less affective disturbance and significantly better performance than low trait denial subjects. Therefore, characteristic users of denial found it served as an effective way to cope with threat in stressful situations. This conclusion is further supported by another study by Houston (1971) where he found that low trait denial subjects were less effective copers in a situation involving threat to self-esteem. In addition, anticipatory cognitive defence mechanisms have been found to decrease levels of arousal under conditions of uncertainty (Averill, 1973).

44. There is sufficient evidence to suggest that cognitive coping can be effective in reducing stress conditions; however, as yet, this work has largely been experimentally confined to laboratory conditions, with little effort to develop a practical strategy for implementing such approaches under real life conditions (Holmes and Houston, 1974).

45. Relaxation techniques coping with stress. Very often stress is an unavoidable part of the person's occupation; for example, managers and air traffic controllers, who are under stress for different reasons, but at the same time derive a great deal of satisfaction from their jobs. Recently, there has been a growing interest in individual relaxation techniques which reduce the tension of a stressful lifestyle, through techniques such as yoga, meditation

and biofeedback. These purport to help the individual "wind down" after a stressful day and, as such, may help to reduce the deleterious effects of occupational life on physical health. Meditation involves restricting awareness to concentrate on one single source or "mantra" (Ornstein, 1972), an activity which is repeated twice a day for approximately 20 minutes. Transcendental meditation (TM) has been reported to help work adjustment through the reduction of tension (Robbins and Fisher, 1972). TM is not only limited to the easing of stress, but also may have other beneficial effects, as Kuna (1975) reviews. It has been found to increase productivity in businessmen, improve attention, increase discriminatory capacity, alertness and reaction time. Frew (1974) concludes that meditators demonstrate increased job satisfaction and performance, less anxiety about promotion, faster work rates, and more organisational authority and responsibility. However, effectiveness of TM largely depends on the orientation and motivation of the individual. Goleman (1976) identifies the importance of meditation in its capacity to break up the threat-arousal-threat-spiral, so that after the individual has experienced a stressful event he can relax himself. He adds that meditators recover more quickly from stress than non-meditators and are better equipped to deal with stress in their lifestyle.

46. Other techniques involve making the individual aware of his body's reaction to stress, whether it be physiological or psychological. Two techniques have been found to help individuals with stressful behaviour to reduce their life stress - through "anxiety management training" and "visuo-motor behaviour rehearsal" (Siunn, 1976). The individual learns the cues which signal the onset of stress, such as tightening of the muscles or increase in heart rate, and he is then taught to reduce this stress through training in muscle relaxation. This has been reported by Suinn (1976) to be effective in lowering stress in people faced with a variety of different personal and occupational problems.

47. A similar technique using yoga and biofeedback, has been used with reasonable success in a preliminary study by Patel (1975). Individuals were taught how to relax and meditate while having continuous information about their progress from a biofeedback instrument, which served to reinforce their responses.

C. Social support and coping with stress

48. A large number of studies have produced evidence indicating the importance of the social group to the

individual as a source of job satisfaction (Cooper, 1975). In brief the "human relations" approach to the workplace emphasises the role of social relationships in achieving a satisfying and rewarding work environment. There is now evidence that the individual's work group and social group may provide effective social support which offsets the effects of stress and CHD (Cooper and Marshall, 1977). In particular, a study of social stress in Japan reveals marked differences in rates of CHD compared to the United States, which appears to be related to certain features of Japanese lifestyle (Matsumoto, 1970). In 1962 the ratio of death rate from CHD to the total death rate was reported to be 33.2 for United States whites and 8.7 for the Japanese (Luisada, 1962). Matsumoto reports that two major factors seem to be increasingly implicated in the development of CHD, namely high fat diet and emotional stress. The diet hypothesis is strongly supported by the fact that the diet of Japan derives less than 16 per cent of its calories from fat as compared with 40 per cent in the American diet. Also, by the findings relative to the difference between Japanese living on different diets (Marmot and Winklestein, 1975), social stress has been found to be related to CHD and is associated with the elevation of serum cholesterol. Matsumoto reports these findings from other studies on medical students during exams (Dreyfuss and Czackes, 1959), accountants during tax preparation deadlines (Friedman et al., 1958), and patients undergoing surgery. Matsumoto hypothesises that the in-group work community of the individual in Japan, with its institutional stress-reducing strategies, plays an important role in decreasing the frequency of disease - "the deleterious circumstances of life need not be expressed in malfunctioning of the physiological or psychological systems if a meaningful social group is available through which the individual can derive emotional support and understanding". Indeed, Japan has a cultural norm towards strong group dependence and, although it has a highly structured society, it is able to counterbalance deleterious stress through effective social support.

D. Preventative approaches to health-related behaviours and CHD

49. As we have seen, certain health-related behaviours and specific types of behaviour patterns give rise to CHD risk. Therefore, in order to reduce the risk engendered by these behaviours, we need to adopt a preventative approach based on behaviour modification. Ball and Turner (1975), referring to CHD, say that "since it is largely caused by man's way of life it is only by behavioural change that the number of deaths can be reduced". However, Henderson and Enelow (1976) point out that changes in

behaviour "must be viewed in relationship to the social structures in which they operate". Preventative measures in this area have been characterised by two main approaches - first, through the modification of single risk factors based on an individual counselling approach,,and second, through the modification of one or more risk factors based on a mass education approach.

50. Recommendations made by the Joint Working Party (1976) refer to dietary changes by reducing the present amount of fat in the diet from over 40 per cent of total calories down to 35 per cent. This could be achieved through a general reduction of saturated fat intake by partially substituting polyunsaturated fats, fruit and fresh vegetables. The modification of eating habits could be achieved through education via the family doctor, or clinic staff, possibly in conjunction with food and diet questionnaires to monitor eating habits (Blackburn, 1975).

51. While diet may be more readily modified, smoking presents more of a problem to people who have become addicted. Although there is no doubt about the role of smoking as a CHD risk factor, there does not appear to be an effective way of inducing cessation of smoking. The Joint Working Party recommends the "advice of a doctor to his patient" as one of the more effective methods of persuasion. Blackburn (1975) also advocates this approach in conjunction with audio-visual aids, group meetings, aversive conditioning and relaxation techniques. The hypothesis that smoking is necessarily a contributor to CHD does not go unchallenged and Seltzer (1968) points to several studies which have found no association between smoking and CHD (e.g. Kozarevic et al., 1971). Selzer (1975) claims that many other factors have to be taken into account, in particular the "constitutional" hypothesis, that is, that coronary-prone individuals "self-select" themselves into smoking. Also, in studies where people have given up smoking, a number of habits may have been modified at the same time which have independent effects on CHD incidence.

52. The evidence regarding physical exercise is rather more sparse. An extensive review of the literature concerning the role of physical activity in the prevention and therapy of CHD concludes that different manifestations of CHD have varying associations with physical activity; however, it finds sufficient support for the opinion that increased physical activity should be propagated (Paul, 1963). Physical exercise is recommended elsewhere (Stamler and Epstien, 1972) - first as a protective measure against a sedentary lifestyle, and second, as a therapeutic measure.

53. The second approach based on mass education has involved a number of multifactorial studies which have been carried out with the aim of controlling one or more risk factors and measuring the effects over time to determine whether there has been any significant reduction in CHD risk as a result of screening and intervention. The current WHO (1974) trials have been set up, first to estimate the extent to which the main risk factors of CHD can be modified in industrial workers using primarily a health education approach and a realistic level of resources; and second, to estimate the effects of such changes on CHD incidence, and the consistency of results between different countries. The study has been set up in four countries with screening and intervention in the form of treatment and advice on cholesterol lowering diet, cessation of smoking, daily physical exercise, weight reduction and hypotensive drug therapy. Certain drawbacks have been identified which can be generally applied to the multifactorial approach. These are: (1) In a very large study with limited resources, adherence to advice tends to be poorer than in a small intensively staffed pilot study. (2) This particular study has only been going for a short time and it remains to be seen whether enthusiasm can be maintained over a period of years. (3) The design is such that the effect of the programme can only be assessed in the intervention community taken as a whole with an inevitable dilution of results. Ball and Turner (1975) also express scepticism as to the success of multifactorial approaches, saying that where several risk factors are present, trials attempting to alter any one factor are likely to give inconclusive results. Criticism has been made about the massive expenditures on the large-scale intervention trials to the comparative exclusion of further research in the etiology of CHD (Marmot and Winklestein, 1975). Furthermore, they claim that even if the intervention programmes are successful, CHD will still remain a "major killer of epidemic proportions" and they conclude that funds ought to be diverted towards further research into the determinants of CHD. Several large-scale intervention studies have been carried out with varying degrees of success and their prospects are reviewed by Stanler and Berkson (1972) who emphasise the need for "second generation trials" based on earlier studies.

Note:

[1] Professor of Organisational Psychology, Department of Management Sciences, University of Manchester Institute of Science and Technology (UMIST), Manchester, United Kingdom.

References

J.R. Averill: "Personal control over aversive stimuli and its relationship to stress", in Psychological Bulletin (Washington, American Psychological Association), 1973, Vol. 80, pp. 286-303.

C.R. Bainton and D.R. Peterson: "Deaths from coronary heart disease in persons fifty years of age and younger: a community-wide study", in New England Journal of Medicine (Boston, Massachussetss Medical Society), 1963, Vol. 268, pp. 569-574.

C.B. Bakker: "Psychological factors in angina pectoris", in Psychosomatics (Greenwich, Cliggott Publishing Co.), 1967, Vol. 8, pp. 43-49.

K.P. Ball and R. Turner: "Realism in the prevention of CHD", in Preventive Medicine (New York, Academic Press Inc.), 1975, Vol. 4, pp. 390-397.

R.T. Beattie, T.G. Darlington and D.M. Cripps: The Management Threshold, BIM Paper OPN 11, 1974.

D. Berkson: "Socio-economic correlates of atherosclerotic and hypertensive heart disease in culture, society and health", in Annals of the New York Academy of Sciences , 1960, Vol. 84, pp. 835-850.

H. Blackburn: "Coronary risk factors: how to evaluate and manage them", in European Journal of Cardiology (Amsterdam, Elsevier North-Holland Biomedical Press), 1975, Vol. 2/3, pp. 249-283.

L. Breslow and P. Buell: "Mortality from coronary heart disease and physical activity of work in California", in Journal of Chronic Diseases (New York, Pergamon Press), 1960, Vol 11, pp. 615-626.

G.W. Brooks and E.F. Mueller: "Serum urate concentrations among university professors", in Journal of the American Medical Association, 1966, Vol. 195, pp. 415-418

R.D.Caplan, S. Cobb and J.R.P. French et al.: Job demands and workers' health (Washington DC, US Department of Health, Education and Welfare Publication No. (NIOSH) 75-160, US Government Printing Office, 1975).

N. Cherry: "Stress, anxiety and work", in Journal of Occupational Psychology (England, British Psychological Society), 1978, Vol. 51, pp. 259-270.

S. Cobb and R.H. Rose: "Hypertension, peptic ulcer and diabetes in air traffic controllers", in Journal of Australian Medical Association, 1973, Vol. 224, pp. 489-492.

C.L. Cooper: Theories of Group Processes (New York and London, John Wiley and Sons, 1975).

C.L. Cooper: Stress Research: Issues for the 80s (New York and London, John Wiley & Sons, 1983).

C.L. Cooper: "Personality characteristics of successful bomb disposal experts", in Journal of Occupational Medicine (Chicago, Flournoy Publishers, Inc.), 1982, Vol. 24, No. 9, pp. 653-655.

C.L. Cooper and M.J. Davidson: High Pressure: Working Lives of Women Managers (Fontana paperback), 1982.

C.L. Cooper and J. Marshall: "Occupational sources of stress: a review of the literature relating to coronary heart disease and mental ill health", in Journal of Occupational Psychology (England, British Psychological Society), 1976, Vol. 49, pp. 11-28.

C.L. Cooper and J. Marshall: Understanding Executive Stress (London, Macmillan, 1977).

C.L. Cooper and A. Melhuish: "Occupational stress and managers", in Journal of Occupational Medicine (Chicago, Flournoy Publishers), 1980, Vol. 22, No. 9, pp. 588-592.

T. Cox: "Repetitive work", in C.L. Cooper and R. Payne (eds): Current Concerns in Occupational Stress (Chichester, New York, John Wiley & Sons, 1980).

M.J. Davidson and A. Veno: "Stress and the policeman", in C.L. Cooper and J. Marshall (eds): White Collar and Professional Stress (London, John Wiley & Sons, 1980).

F. Dreyfuss and J.W. Czackes: "Blood cholesterol and uric acid of health medical students under stress of examination", in Archives of Internal Medicine, (Chicago, American Medical Association), 1959, Vol. 103, p. 708.

M.T. Eaton: "The mental health of the older executive", in Geriatrics, 1969, Vol. 24, pp. 126-134.

J.M. Erickson, W.M. Pugh and K.E. Gunderson: "Status congruency as a prediction of job satisfaction and life stress", in Journal of Applied Psychology (Washington,

American Psychological Association), 1972, Vol. 56, pp. 523-525.

The Framingham Study: "An epidemiological investigation of cardiovascular disease", in W.B. Karrel and T. Gordon (eds): US Public Health Service, 1970-71, Sections 26 and 27.

M. Friedman: Pathogenesis of Coronary Artery Disease (New York, McGraw-Hill, 1969).

M. Friedman, R.H. Rosenman and V. Carroll: "Changes in serum cholesterol and blood clotting time in men subjected to cyclic variations of occupational stress", in Circulation (Texas, American Heart Association), 1958, Vol. 17, pp. 852-861.

M. Friedman and R.H. Rosenman: Type A Behaviour and Your Heart (New Haven, Conn., Fawcett Publications, 1974).

J.R.P. French and R.D. Caplan: "Psychosocial factors in coronary heart disease", in Industrial Medicine, 1970, Vol. 39, pp. 383-397.

J.R.P. French and R.D. Caplan: "Organisational stress and individual strain", in A.J. Marrow (ed): The Failure of Success (New York, Amacon, 1972), pp. 31-66.

J.R.P. French, C.J. Tupper and E.I. Mueller: "Workload of university professors" (Ann Harbor, Mich., University of Michigan, 1965). [Unpublished research report].

D.R. Frew: "Transcendental meditation", in Academic Management Journal, 1974, pp. 362-368.

D. Goleman: "Medication helps break the stress spiral", in Psychology Today (New York, Ziff Davies Publishing Co.), 1976.

J.B. Henderson and A.J. Enelow: "The coronary risk factor problem: a behavioral perspective", in Preventive Medicine (New York, Academic Press), 1976, Vol. 5, pp. 128-148.

L.E. Hinkle: "The concept of 'stress' in the biological and social sciences", in Science, Medicine and Man, 1973, Vol. 1, pp 31-48.

D.S. Holmes and B.K. Houston: "Effectiveness of situation redefinition and affective isolation in coping with stress", in Journal of Personnel and Social Psychology, 1974, pp. 212-218.

B.K. Houston: "Trait and situational denial and performance in stress", in Journal of Personnel and Social Psychology, 1971, Vol. 18, pp. 289-293.

B.K. Houston: "Viability of coping strategies, denial and response to stress", in Journal of Personnel and Social Psychology, 1973, Vol. 41, pp. 50-58.

C.D. Jenkins: "Psychologic and social precursors to coronary disease", in New England Journal of Medicine (Boston, Massachussetts Medical Society), 1971, Vol. 284 (5), pp. 244-255.

C.D. Jenkins: "Psychologic and social precursors to coronary disease", in New England Journal of Medicine (Boston, Massachussetts Medical Society), 1971 b, Vol. 284 (6), pp. 307-317.

S.V. Kasl: "Mental health and work environment: an examination of the evidence", in Journal of Occupational Medicine (Chicago, Flournoy Publishers), 1973, Vol. 15, pp. 506-515.

M. Kelly and C.L. Cooper: "Stress among blue collar workers", in Employee Relations (England, CB Publications, Ltd.), 1981, pp. 2, 3, 6-9.

A. Keys: "Coronary heart disease in seven countries", in Circulation (Texas, American Heart Association), 1970, Vol. 41.

D. Kozarevic, B. Pirc, T.R. Dawber et al.: "Prevalence and incidence of coronary disease in a population study, the Yugoslavia cardiovascular disease study", in Journal of Chronic Diseases (New York, Pergamon Press), 1971, Vol. 24, p. 495.

S.P. Kritsikis, A.L. Heinemann and S. Eitner: "Die angina pectoris im aspekt ihrer korrelation mit biologischer disposition psychologischen und soziologischen emfluss-faktoren", in Deutsch Gesundheit, 1968, Vol. 23, pp. 1878-1885.

D.J. Kuna: "Meditation and work", in Vocational Guidance Quarterly (Leesburg, Va, American Personnel and Guidance Association), 1975, Vol. 2, pp. 70-78.

R. Lazarus: Psychological stress and the coping process (New York, McGraw-Hill, 1967).

R. Lazarus, E. Opton, M. Nomikos and N. Rankin: "The principle of short circuiting of threat: further evidence", in Journal of Personnel and Social Psychology, 1965, Vol. 33, pp. 622-635.

L.H. Lofquist and R.V. Dawis: Adjustment of Work (New York, Appleton Century Crofts, 1969).

A.A. Luisada: "Introduction of symposium on the epidemiology of heart disease", in American Journal of Cardiology, (New York, Technical Publishing Co.), 1962, Vol. 10, p. 316.

G. Mandler and D.L. Watson: "Anxiety and the interruption of behaviour", in C.D. Spielberger (ed): Anxiety and Behaviour (New York, Academic Press), 1966.

S. Marcson: Automation, Alienation and Anomie (New York, Harper and Row, 1970).

B.L. Margolis, W.H. Kroes and R.P. Quinn: "Job stress: an unlisted occupational hazard", in Journal of Occupational Medicine (Chicago, Flournoy Publishing), 1974, Vol. 16, pp. 654-661.

R.U. Marks: "Social stress and cardiovascular disease", in Milbank Memorial Fund Quarterly (Cambridge, MIT Press), 1967, Vol. 65, pp. 51-107.

M. Marmot and W.J. Winkelstein: "Epidemiologic observations on intervention trials for prevention of coronary heart disease", in American Journal of Epidemiology (Baltimore, Society for Epidemiologic Research), 1975, p. 101.

Y.S. Matsumoto: "Social stress and coronary heart disease in Japan", in Milbank Memorial Fund Quarterly (Cambridge, MIT Press), 1970, Vol. 48.

R.R. McCrae, P.T. Costa and R. Bosse: "Anxiety, extraversion and smoking", in British Journal of Social and Clinical Psychology (England, British Psychological Society), 1978, Vol. 17, pp. 269-273.

A.J. McMichael: "Personality, behavioural and situational modifiers of work stressors", in C.L. Cooper and R. Payne (eds): Stress at work (Chichester, New York, John Wiley & Sons, 1978), pp. 127-147.

R.N. McMurray: "Mental illness: society's and industry's six billion dollar burden", in R.L. Noland (ed): Industrial Mental Health and Employee Counselling (New York, Behavioural Publications, 1973).

R.E. Ornstein: The Psychology of Consciousness (New York, Viking Press, 1972).

H.J. Otway and R. Misenta: "The determinants of operator preparedness for emergency situations in nuclear power plants". Paper presented at Workshop on Procedural and Organisational Measures for Accident Management: Nuclear Reactors International Institute for Applied Systems Analysis, Laxenburg, Austria, 28-31 January 1980.

J.M. Pahl and R.E. Pahl: Managers and Their Wives (London, Allen Lane, 1981).

R.S. Paffenbarger, P.A. Wolf and J. Notkin: "Chronic disease in former college students", in American Journal of Epidemiology (Baltimore, Society for Epidemiological Research), 1966, Vol. 83, pp. 314-328.

C. Patel: "Yoga and biofeedback on the management of hypertension", in Journal of Psychosomatic Research (New York, Pergamon Press), 1975, Vol. 19.

O. Paul: "A longitudinal study of coronary heart disease", in Circulation (Texas, American Heart Association), 1963, Vol. 28, pp. 20-31.

R. Payne: "Organisational stress and social support", in C.L. Cooper and R. Payne (eds): Current Concerns in Occupational Stress (Chichester, New York, John Wiley & Sons, 1980), pp. 269-298.

S. Pell and C.A. D'Alonzo: "Myocardial infarcation in a one year industrial study , in Journal of American Medical Association (Chicago, American Medical Association), 1958, Vol. 166, pp. 332-337.

G. Pincherle: "Fitness for work", in Proceedings of the Royal Society of Medicine (London, Royal Society of Medicine), 1972, Vol. 65, pp. 321-324.

C.B. Quinlan, J.G. Burrow and C.G. Hayes: The association of risk factors and CHD in Trappist and Benedictine monks. Paper presented to the American Heart Association, New Orleans, 1969.

Report of the Joint Working Party of the Royal College of Physicians of London and the British Cardiac Society, in Journal of the Royal College of Physicians (London, Update Publications, Ltd.), 1976. Vol. 10.

J. Robbins and D. Fisher: "Tranquility Without Pills" (New York, Peter H. Wyder, 1972).

R.H. Rosenman, M. Friedman and C.D. Jenkins: "Clinically unrecognised myocardial infarcation in the Western collaborative group study", in American Journal of Cardiology (New York, Technical Publishing Co.), 1976, Vol. 19, pp. 776-782.

R.H. Rosenman, M. Friedman and R. Strausse: "A predictive study of CHD", in Journal of the American Medical Association (Chicago, American Medical Association), 1964, Vol. 189, pp. 15-22.

R.H. Rosenman, M. Friedman and R. Strauss: "CHD in the Western collaborative group study", in Journal of the American Medical Association (Chicago, American Medical Association), 1966, Vol. 195, pp. 86-92.

H.I. Russek and B.L. Zohman: "Relative significance of hereditary diet and occupational stress in CHD of young adults", in American Journal of Medical Science, 1958, Vol. 235, pp. 266-275.

S.B. Sells: "On the nature of stress", in J.E. McGrath (ed): Social and Psychological Factors in Stress (New York, Holt, Rinehart and Winston, 1970).

C.C. Seltzer: "The effect of smoking on coronary heart disease", in Journal of the American Medical Association (Chicago, American Medical Association), 1968, Vol. 203, p. 193.

C.C. Seltzer: "Smoking and cardiovascular disease", in American Heart Journal (St. Louis, CV Mosby Co.), 1975, p. 90.

H. Selye: "The general adaptation syndrome and the disease of adaptation", in Journal of Clinical Endocrinology (England, Journal of Endocrinology Ltd.), 1946, Vol. 6, p. 117.

H. Selye: Stress in Health and Disease (Boston, London, Butterworths, 1976).

J.M. Shepard: Automation and Alienation (Cambridge, Mass., MIT Press, 1971).

A. Shirom, D. Eden, S. Wilberwasser and J.J. Kellerman: "Job stress and risk factors in coronary heart disease among occupational categories in Kibbutzim", in Social Science and Medicine (New York, Pergamon Press), 1973, Vol. 7, pp. 875-892.

D.M. Spain: "Problems in the study of coronary atherosclerosis in population groups in culture, society and health", in Annals of the New York Academy of Sciences, 1960, Vol. 84, pp. 816-834.

J. Stamler and D.M. Berkson: "Prospects and multifactorial approaches emphasising improvements in lifestyle", in Advances in Experimental Medicine and Biology, 1972, Vol. 26, pp. 213-244.

J. Stamler and E. Epstein: "Coronary heart disease: risk factors as guides to preventive action", in Preventive Medicine (New York, Academic Press), 1972, Vol. 1, pp. 27-48.

J. Stamler, M. Kjeslsberg and Y. Hall: "Epidemiologic studies of cardiovascular-renal diseases: I. Analysis of mortality by age-race-sex-occupation", in Journal of Chronic Diseases (New York, Pergamon Press), 1960, Vol. 12, pp. 440-455.

R.M. Suinn: "How to break the vicious cycle of stress", in Psychology Today (New York, Ziff Davies Publishing Co.), 1976.

S.L. Syme, M.M. Hyman and P.E. Enterline: "Some social and cultural factors associated with the occurrence of coronary heart disease", in Journal of Chronic Diseases (New York, Pergamon Press), 1964, Vol. 17, pp. 277-289.

W.B. Tethune: "Emotional problems of executives in time", in Industrial Medicine and Surgery, 1963, Vol. 32, pp. 1-67.

W.I. Wardwell, M.M. Hyman and C.B. Bahnson: "Stress and coronary disease in three field studies", in Journal of Chronic Diseases (New York, Pergamon Press), 1964, Vol. 17, pp. 73-84.

A.T. Welford: "Stress and performance", in Man Under Stress (London, Francis and Taylor, 1974).

WHO European Collaborative Group: "An international controlled trial in the multi-factorial prevention of coronary heart disease", in International Journal of Epidemiology (Baltimore, Society for Epidemiological Research), 1974, p. 3.

APPENDIX

LIST OF PARTICIPANTS

EXPERTS

Mr. L.D. BELLMER,
Director,
Union Relations,
Westinghouse Electric Corp.,
Gateway Center,
Pittsburgh, PA.
(United States)
(Employer)

Mr. R. BLASING,
Bundervereinigung der Deutschen
Arbeitgeberverbände (Köln),
Talstrasse 8,
7016 Gerlingen.
(Federal Republic of Germany)
(Employer)

Mr. E.A.M. Van den BOSCH,
Hoogovens Groep BV,
Postbus 10000,
1970 CA Ijmuiden.
(Netherlands)
(Employer)

Dra. M. de F. CANTIDIO MOTA,
Coordinadore del Servicio de Higiene
y Seguridad Industrial,
Servicio Social Industrial,
Confederaçao Nacional da Industria,
735 - 10° andar,
20.030 Rio de Janeiro.
(Brazil)
(Employer)

Mr. L. CORTEBEECK,
Service Entreprise et Secrétaire de la
Fédération de Malines,
Confédération des Syndicats chrétiens
de Belgique,
Rue de la Loi, 121,
1040 Brussels.
(Belgium)
(Worker)

Mr. N. KAMYAMA,
Manager, Second Section,
Employment Development Division,
National Institute of Employment and
Vocational Research,
1-1, 4 chome, Nakano,
Nakanoku,
Tokyo.
(Japan)
(Government)

Dr. M. MORITZ,
Gewerkschaft der Privatangestellten (OGB),
Ausschuss für Automation und
Arbeitsgestaltung,
Deutschmeisterplatz 2,
1010 Vienna.
(Austria)
(Worker)

Mr. P.L. REMY,
Director,
Agence nationale pour l'amélioration des
conditions de travail (ANACT),
7 bd. Romain Rolland,
92128 Montrouge.
(France)
(Government)

Mr. N.A. SAFRONOV,
Deputy Director,
Research Labour Institute,
Moscow.
(USSR)
(Government)

Mr. A. SCHULTE,
Federal Ministry of Labour and Social
Affairs,
Postfach 140280,
5300 Bonn 1.
(Federal Republic of Germany)
(Government)

Mr. B. SELLES,
Union général des travailleurs
algériens (UGTA),
52A, rue Belouizdad,
Algers.
(Algéria)
(Worker)

Mr. V.G. VERETTENIKOV,
Chef adjoint,
Département des salaires et du travail
économique,
Conseil central des syndicats soviétiques,
42 Leninsky Prospekt,
117119 Moscow
(USSR)
(Worker)

Personal adviser

Mrs. R. KALINKINA,
Département international,
Conseil central des syndicats
soviétiques,
42 Leninsky Prospekt,
117119 Moscow.
(USSR)

OBSERVERS

Mrs. B. FAUCHERE,
World Confederation of Labour,
1, rue de Varembé,
Case postale 122,
1211 Geneva 20 CIC.

Mr. B. ROBEL,
World Confederation of Labour,
1, rue de Varembé,
Case postale 122,
1211 Geneva 20 CIC.

Mr. L. LABRUNE,
Fédération syndicale mondiale (FSM),
10, rue Fendt,
1201 Geneva.

Mr. E. LAURIJSSEN,
Assistant Director of Geneva Office,
International Confederation of Free Trade
Unions (ICFTU),
27, rue Coulouvrenière,
1204 Geneva.

Mr. C. KAPARTIS,
Secrétaire général adjoint,
Organisation internationale des
Employeurs (OIE),
28, chemin de Joinville,
Case postale 68,
1216 Cointrin.

Ms. C. REAVIS,
Organisation internationale des
Employeurs (OIE),
28, chemin de Joinville,
Case postale 68,
1216 Cointrin
